The Age

ALSO OF INTEREST AND FROM MCFARLAND

Edited by Cory Barker, Chris Ryan *and* Myc Wiatrowski
Mapping Smallville: *Critical Essays*
on the Series and Its Characters (2014)

Edited by James F. Iaccino, Cory Barker and Myc Wiatrowski
Arrow *and Superhero Television: Essays on Themes*
and Characters of the Series (2017)

LIBRARY OF CONGRESS CATALOGUING-IN-PUBLICATION DATA

Names: Barker, Cory, 1988– editor. | Wiatrowski, Myc, editor.
Title: The age of Netflix : critical essays on streaming media,
    digital delivery and instant access / edited by Cory Barker
    and Myc Wiatrowski.
Description: Jefferson, North Carolina : McFarland & Company, Inc.,
    Publishers, 2017 | Includes bibliographical references and index.
Identifiers: LCCN 2017030294 | ISBN 9780786497478 (softcover :
    acid free paper) ∞
Subjects: LCSH: Netflix (Firm) | Streaming video—Social aspects. |
    Streaming technology (Telecommunications)—Social aspects. |
    Mass media—Social aspects.
Classification: LCC HD9697.V544 N48235 2017 | DDC 384.55/506573—
    dc23
LC record available at https://lccn.loc.gov/2017030294

BRITISH LIBRARY CATALOGUING DATA ARE AVAILABLE

ISBN (print) 978-0-7864-9747-8
ISBN (ebook) 978-1-4766-3023-6

© 2017 Cory Barker and Myc Wiatrowski. All rights reserved

*No part of this book may be reproduced or transmitted in any form
or by any means, electronic or mechanical, including photocopying
or recording, or by any information storage and retrieval system,
without permission in writing from the publisher.*

Front cover images © 2017 iStock

Printed in the United States of America

*McFarland & Company, Inc., Publishers
    Box 611, Jefferson, North Carolina 28640
    www.mcfarlandpub.com*

The Age of Netflix

*Critical Essays on Streaming
Media, Digital Delivery
and Instant Access*

Edited by Cory Barker
and Myc Wiatrowski

McFarland & Company, Inc., Publishers
Jefferson, North Carolina

Table of Contents

Acknowledgments

The editors would like to thank all 11 contributors for their meticulous work, consistent kindness, and true patience as the collection grew from a compelling idea into a satisfying final product. Their wonderful insights found within these essays made it easy to remain committed to this project amid grueling schedules, job changes, and much more.

Cory would like to thank his coeditor Myc for his work and commitment to bringing a great idea to life. Cory would also like to shout out to his friends in the critical and scholarly community for always lending their time and ears to workshop ideas big and small and his family for their warmth and support.

Myc would like to thank his coeditor, Cory, whose enthusiasm and dedication brought this collection together. Without his hard work, this book would never have come to fruition. Myc also owes his greatest debt to his family, Laura and Lucas, whose endless patience and understanding have made this all possible and worthwhile.

Introduction

CORY BARKER *and* MYC WIATROWSKI

In 2016, citizens in the United States and around the globe were forced
to confront the deep-seeded political, economic, and cultural divisions among
themselves. From shocking voting results in the U.S. presidential election and the United Kingdom's departure from the European Union to heated
debates about "fake news" and the "filter bubbles" of social media to the
(re)emergence of fringe groups driven by nationalism, hatred, and conspir-
atorial thought, the modernized world experienced more tumult than usual.
Nonetheless, despite the very real partitions among people, one entity continues to bring us together—sort of. Already with an enormous footprint in
the United States, Netflix expanded to 130 new countries in early 2016.[1] As
detailed in its third quarterly report in October 2016, the company added
more than 12 million new subscribers in the year's first nine months and
brought its total number of paying customers (and thus not including those
who share accounts or passwords) to nearly eighty-seven million.[2]

The brilliance of Netflix's strategy is in how its streaming video library
manages to appeal to disparate groups of people across the world *without* a
unified cache of content. Indeed, the company takes the opposite approach,
using its sophisticated algorithm and seemingly endless resources to buy,
develop, and distribute as many different types of content to as many micro-
targeted audience groups as possible.[3] This data-driven narrowcasting manifests in a variety of genres of programming tailored to particular audiences,
including the prestige drama (*House of Cards* [2013–] and *Bloodline* [2015–
2017]), the rebooted multi-camera sitcom (*Fuller House* [2015–] and *One Day
at a Time* [2016–]), and the superhero franchise (Marvel's *Daredevil* [2015–],
Jessica Jones [2015–], and *Luke Cage* [2016–]). However, in 2016, Netflix took
this approach even more globally, introducing more non–English language
series such as France's *Marseille* (2016–) and Brazil's *3%* (2016–).

Netflix's worldwide expansion almost guarantees that the company will

1

further integrate itself into our everyday lives. Since its now-famous shift from physical media rentals to a high-definition streaming video platform, Netflix's stature has grown significantly. In North America, the company finds itself at the fulcrum of countless industrial, cultural, economic, technological, and political developments. Its role in the popularization of streaming video has fundamentally altered the ways in which we watch, discuss, and generally consume media. From the rise of binge-watching and password-sharing to intermittent debates about spoiler etiquette and how critics should cover programs that are released all at once, Netflix is the central force in the contemporary experience of media consumption. The company has an equally notable impact on how television and film is produced, distributed, and marketed. Armed with a large operating budget, Netflix has improved its position within Hollywood's inner circle since 2012, outbidding HBO for A-list talent as well as spending lavishly on independent films across the festival circuit. Much of what Amazon or Hulu or even HBO has done in recent years has been in response to Netflix's embrace of original or exclusive content, setting off an arms race to craft the most valuable subscription streaming video service.

Meanwhile, Netflix projects are not only meticulously targeted with audiences' taste profiles in mind, they are also immaculately marketed and "eventized" to cut through modern popular culture's dense clutter. The company's streaming of full seasons all at once situates those releases as must-watch *and* must-complete occurrences—and is a tactic that networks and cable channels have mimicked in recent years.[4] Yet, the existence of the ever-changing Netflix library taps into the phenomenon of the long tail, with consumers always having another new-to-them series or film to watch years after its initial release. As a result, the company manages to imbue its library with a sense of perpetual personalized discovery that, in theory, offers enough content to keep consumers subscribing from month to month and year to year.

Although much of the attention paid to Netflix hinges on its influence on consumers and industry practice, the company is similarly relevant in other arenas. The influx of cord cutting—consumers unsubscribing to traditional cable packages—over the past five years is regularly attributed to streaming video and Netflix more specifically.[5] Cable and Internet companies have tried to lure customers back with bundles *including* Netflix trials, but have more recently turned toward cheaper, more targeted "skinny bundle" options to reach those who, for example, may appreciate the personalized Netflix experience but still want to watch live sports on ESPN.[6] Netflix's embrace of streaming video and subsequent indifference in its DVD rental business has also been identified as a catalyst in the death of physical media, with cultural critics and scholars decrying the declining availability of older and more obscure films and television series. As Hollywood and consumers move toward Netflix and further away from physical media, more products fail to make the jump to the next platform, whether due to rights issues, conversion challenges, or perceived lack of demand.[7]

Netflix has also been a key figure in the discourse surrounding net neutrality and data caps, perhaps most notably when it reached an agreement with Comcast to ensure that subscribers would receive Netflix content at "faster and more reliable speeds."[8] In the aftermath of the deal, Netflix executives repeatedly went on the record in support of net neutrality.[9] Conversely, the company has installed a ban on customers trying to watch via a VPN (virtual private network) as part of a larger process known as "geo-blocking," wherein content is not accessible to those outside a particular geographic location.[10] Meanwhile, American cities like Pasadena, California, have proposed a so-called "Netflix tax" to recoup lost revenues from increased cord-cutting, while Chicago has successfully installed its own "cloud tax" on streaming services like Netflix that are "delivered electronically."[11]

Altogether, these efforts illustrate the prominence of Netflix beyond binge-watching and all-at-once release strategies. Both Netflix and its opponents within the government have displayed a predictably inconsistent perspective on who can access its streaming library, and what those people should be required to do—or, perhaps better said, how much they should be required to pay—to make that access possible. Likewise, as a technology company driven by the contemporary Silicon Valley ethos of "get big fast," Netflix has been less concerned about what content is left behind in the march toward the great streaming video singularity.

Although these headline-grabbing data points and anecdotes underline Netflix's disruptions of culture, less discussed is the company's uncanny ability to build on pre-existing business models or industry practices. The DVD rental service of course combined the video store with Amazon's nationwide shipping practices. The move to streaming video followed both Apple's iTunes store and similar streaming platforms developed by U.S. and UK broadcasters. The shift from licensed content to original products mirrored the path traveled by countless American cable channels, from HBO to TNT to MTV. These realities do not limit Netflix's centrality to modern life, but simply serve as a reminder that, as Lisa Gitelman asserts, "media are themselves denizens of the past. Even the newest media today come from somewhere, whether that somewhere gets described broadly as a matter of supervening social necessity, or narrowly in reference to some proverbial drawing board and a round or two of beta testing."[12] Thus, while this collection investigates why Netflix is omnipresent in our lives, it also returns to how the company evokes the phenomena of years past and of other cultures. The book is focused on what Gitelman calls protocols, or the "vast clutter of normative rules and default conditions, which gather and adhere like a nebulous array around a techno-logical nucleus," that have been altered, replaced, or introduced with the advent of this streaming platform.[13] Still, authors resist the urge to proclaim Netflix as a singular power among changing protocols.

With this in mind, this collection consists of three parts that aspire to underline the far-reaching but occasionally contradictory influence of Netflix around the globe. Each part addresses how Netflix's attempts to be many things—technology company, media conglomerate, policy stakeholder, and so on—and generate products and protocols that consumers have both cel-ebrated and decried. In Part One, "Netflix as Disruptor and as Cultural Insti-tution," the authors highlight three core developments that are the result of Netflix's recent success. In the first essay of this section, Justin Grandinetti analyzes the company's all-at-once release model and its effect on consumers but also critics, who have been confronted with the challenge of writing about an entire season of television in a weekend or risk losing the web traffic that television coverage generates. In this realm, Netflix has functioned as a par-ticularly disruptive force, destabilizing the popular—and, for digital media companies, lucrative—style of television criticism, wherein writers produce recaps and deconstructions of episodes that subsequently enable viewers to discuss the episode among themselves in the comments section. Websites like *Television Without Pity* and *The A.V. Club* popularized this style, repli-cating a more obsessive form of water cooler conversation in a digital envi-ronment. While viewers have been navigating the challenges of these digital water coolers for years—including but not limited to spoiler etiquette and disputing factions within fandom—the release and coverage of a full season expedites those challenges for readers and critics alike. In spotlighting com-mentary from both groups in the early stages of Netflix's all-at-once rollout, Grandinetti shows how the contours of communal television have been rede-fined on the fly, with everyone caught between the traditional television dis-tribution schedule and the desire to participate in the online conversation.

Djoymi Baker concentrates on another essential characteristic of Netflix: binge-watching. However, rather than detailing how consumers have embraced this kind of viewing practice (as others later in the collection do), Baker turns his attention to how the company frames its original programming *as* binge-worthy. For Netflix, the simultaneous release of a season's episodes is only part of the strategy to inspire binge-viewing. Many of its original series eschew the discrete episodic structure of television, with fewer standalone episodes and more collections of scenes moving slowly to a climax. Netflix also utilizes A-list casts, lavish budgets, and endless marketing resources to brand its original productions as must-see events. Netflix is not just encour-aging binge-viewing; the company crafts what Baker, borrowing from Vivian Sobchack, calls "epic-viewing." For Baker, Netflix's approach recalls not only

HBO's cunningly marketed "Not TV" era, but also the epic miniseries of 1970s and 1980s broadcast television. Examining the company's production and promotional practices, Baker underlines how Netflix elongates Sobchack's "excess of temporality," where television is one exhaustive, epic text.

The last essay of Part One moves to Netflix's position within the net neutrality debates. Using Netflix's controversial agreement with Comcast as a starting point, Joseph Donica examines the former's broader role in the current structure—and potential future—of the Internet. Netflix has aligned with "open Internet" supporters, positioning itself as a kind of "good" corporation that has its eyes on more than the bottom line. Nevertheless, Netflix surely spoke out for net neutrality to protect its place within the market, as more vertically integrated competitors stood to benefit from less restrictive regulations. These contradictory positions are at the center of Donica's analysis, where Netflix is both an open Internet advocate with a "hacker ethic" corporate culture and the target of criticisms from leftist groups expecting more from a supposedly outspoken corporation. Donica details how Netflix unintentionally became a central player in the net neutrality controversy, and occasionally acts in a manner that illustrates Silicon Valley's murky understanding of freedom and capitalism.

The collection's second part, "Netflix as Producer and as Distributor," concentrates on a trio of vital Netflix projects: the original series *Orange Is the New Black*, the revival season of Fox's *Arrested Development* (2013), and the Sundance Film Festival documentary *The Square* (2014). In this part, the authors consider how the presence of Netflix as a partner transforms the processes of production, distribution, and reception. Like Grandinetti and Baker, Maria San Filippo displays an interest in how Netflix's distribution model affects the temporal and communal experiences of television. Unlike the premiere of *House of Cards* or *Arrested Development*'s big return, the first season of *OITNB* arrived on Netflix in 2013 with relatively little hype, only to generate the most passionate response of all the company's original series. Acclaim for the series stemmed from its commitment to representational diversity and its transgressive spirit within the framework of a prestige dramedy. San Filippo considers how queer viewers, commonly ignored by the machinery of mainstream Hollywood, are both freed and limited by *OITNB* and its all-at-once release. Noting the wave of passionate online conversation that comes with a Netflix release, San Filippo argues that queer viewers lose some of the "embodied interactions" that are central to their subcultural identities. Still, in her analysis of character arcs and storylines, San Filippo shows that the series nevertheless offers multifaceted expressions of queerness *and* a queering of television's structures of temporality and seriality.

The revival of cult Fox sitcom *Arrested Development* was one of the first signs that Netflix was serious about its shift from distributor to producer.

Although the Mitch Hurwitz–created project upended some of the sitcom's fundamental conventions during its Fox run, Hurwitz and the creative team (including producers Ron Howard and Brian Grazer) used the freedom granted to them by Netflix to take even more risks with the return season. These risks are at the center of the essay by Maíra Bianchini dos Santos and Maria Carmem Jacob de Souza. Beginning with the history of *Arrested Development*'s journey to Fox's schedule, the authors trace how Hurwitz and company developed the series' verité style and later transformed a self-conscious and layered vision into an even more fragmented, referential puzzle. The authors show how prior projects (Howard's work on the reality television-esque *EdTV* [1999]) and production obstacles (scheduling challenges with the series' large cast) regularly inspired the experimentation on all seasons of *Arrested Development*. Thus, Netflix is shown to be a respectful partner, but not the singular power crafting one of television's most unique projects.

Though most of Netflix's recent successes are related to its role as a producer of television, the company has begun to expand its footprint as a distributor of first-run films as well. Again, a vital component of Netflix's strategy is borrowed from normal Hollywood practice. The company has become a mainstay at the world's major film festivals, including at Sundance, where the documentary *The Square* first debuted in January 2013. An Academy Award nominee in 2014, *The Square* depicts the Egyptian Revolution beginning with the 2011 uprising in Tahrir Square. James N. Gilmore's essay considers how the circulation of *The Square*'s sharply political messages on a digital platform like Netflix speak to the potential power of what Henry Jenkins, Joshua Green, and Sam Ford call "media spreadability." In analyzing the film's do-it-yourself (DIY) production aesthetics and calls to action, Gilmore asserts that *The Square* is tailor-made for distribution across the web and social media. Netflix, he posits, is equally positioned to circulate political documentaries around the globe, dramatically expanding the civic influence of film.

The final part of the collection, "Netflix as Narrowcaster and as Global Player," looks at Netflix's impact on both the small and large scale. The first two essays examine the Netflix-driven trends of binge-watching and algorithmic recommendations through the lens of viewer response. Both essays provide observations from extensive interviews with and journaling from Netflix users who speak to the ways in which the company's practices have altered their everyday lives. Emil Steiner reports on binge-watching, the phenomenon most associated with Netflix's rise to prominence. Drawing a line from television's status as a "boob tube" and "vast wasteland" to the stigmas surrounding binge-viewing, Steiner elucidates how the medium has long been at the center of critiques against popular culture. Participants in Steiner's study regularly embody and eschew cultural assessments of bingeing, acknowledging how the practice reformulates their viewing habits, attentiveness, and overall relationship with programming. However, as Steiner and his interviewees traverse the touchy subject of how much bingeing is too much, it becomes clear that, as with all viewing practices, there is no unified binge-viewing experience. Netflix users see bingeing as something to celebrate and something to be ashamed of; likewise, they binge-watch to catch up, to veg out, to re-watch, and to share.

While Steiner traces Netflix's place within historical debates about accepted viewing practices, Alison N. Novak investigates how the company's data-driven approach compares to previous generations of narrowcasting. Like retail brands, magazine publishers, and cable channels before them, Netflix narrowcasts to its multitudinous user base. And like Amazon and Google, the company relies heavily on data collection and sophisticated algorithms to produce personalized recommendations to different user segments. Narrowcasting, recommendation engines, and algorithms are too often deployed as buzzy signifiers for the emergent digital ecosystem, but Novak's findings underline how they actually operate in practice. Novak narrowcasts her own research, surveying the millennial response to Netflix's recommendations. Though the most coveted demographic, Novak displays how millennials are also quite aware of how data collection affects everything they consume. On one hand, participants in Novak's study underline the gradual banality of personalized recommendations, as most of the millennial respondents express comfort with and appreciation for how Netflix cultivates their taste profiles. On the other hand, these participants affirm that Netflix's practices are part of a pattern of narrowcasting and technology, one that they are sure will lead to more nefarious forms of artificial intelligence and computer learning.

The remaining essays of this part reflect on Netflix's mounting influence outside of the United States. Here again, authors position Netflix not as a supreme innovator but instead as a key player within broader technological and industrial trends in the United Kingdom and Mexico. In the UK television industry, for instance, Netflix followed Sky Digital, a groundbreaking interactive satellite digital television service, to the marketplace. Having previously researched consumer response to the rollout of Sky Digital in the mid–2000s, Vivi Theodoropoulou is well equipped to compare how Netflix has mirrored and expanded Sky's playbook for a new generation. Theodoropoulou juxtaposes past and present interviews to exhibit how habits or phenomena we generally associate with Netflix have existed long before its streaming platform took hold. The interviews also reveal a crucial point of contention with video libraries of all kinds: the availability of content. Sky Digital and Netflix promise users the ability to watch what and when they want, but Theodoropoulou shows how both generations are still restricted by what is available at any given time. For Netflix interviewees, there is a firm understanding, built through experience, that access to content is always the result of nego-tiations beyond their control. As such, Theodoropoulou argues that it is the content—and not the platform—that will continue to persevere across each new technological development.

The final essay presents a vision of Netflix's possible futures in its era of global expansion. Before its sprawling growth campaign in 2016, Netflix moved much more cautiously into other territories, including Mexico. Elia Margarita Cornelio-Marí surveys Netflix's 2011 entry into Mexico with a par-ticular focus on the industrial, regulatory, and cultural challenges that the company faced. Cornelio-Marí shows that, while Netflix identified Mexico as a target for enormous potential revenue, the streaming giant immediately encountered an audience hungry for local content and sympathetic toward piracy. The essay spotlights a Netflix that is quick to adapt its practices to non–American environments, as Cornelio-Marí details the company's tin-kering with pricing, crafting of new marketing strategies, dubbing of foreign content, and cultivating of localized productions. However, Netflix's arrival in any new location is not just a story about Netflix or its new customers; it is also about the response from local competitors and legislators. To this end, Cornelio-Marí shows how vertically and horizontally integrated corporations have begun to use their vast infrastructural resources and local knowledge to raise the level of competition against Netflix. Ultimately, the Mexico case study demonstrates that for Netflix, worldwide expansion will not be easy. The company must strike a balance between the local and the global—a chal-lenge that only gets harder with each additional expansion.

With a subject like Netflix, the talking points are endless; this collection seeks only to participate in the ongoing discourse about Netflix's place in contemporary culture. Still, these ten essays establish that to best understand an innovative company like Netflix, we must continue to look backward, toward its antecedents in the technology and media industries. There we are reminded that sweeping changes in industry practice, consumer habits, and policy are rare, and more commonly part of a granular trek toward the future. There we are also reminded that while Netflix appears to have solidified its place—as a disruptor and cultural institution, as a producer and distributor, and as a narrowcaster and global player—the future it faces is still uncertain.

NOTES

1. Nathan McAlone, "Netflix Just Launched in 130 More Countries, and the Stock Is Soaring," *Business Insider*, January 6, 2016, accessed January 20, 2017, http://www.businessin sider.com/netflix-just-launched-in-130-countries-more-countries-2016-1.

2. Jeff Dunn, "Netflix Is Booming on the Back of Subscribers Outside of the U.S.," *Business Insider*, October 18, 2016, accessed January 20, 2017, http://www.businessinsider.com/net flix-subscribers-us-international-chart-2016-10.

3. According to the October 2016 quarterly report, Netflix vowed to spend more than $6 billion on its original series productions in 2017. See Lauren Gensler, "Netflix Is Still

Spending Money Like There's No Tomorrow," *Forbes*, October 18, 2016, accessed January 20, 2017, http://www.forbes.com/sites/laurengensler/2016/10/18/netflix-cash-flow-original-content/#6fdd970a4f52.

4. Cynthia Littleton, "NBC Embraces Binge-Viewing, Releasing All 'Aquarius' Episodes Online After Premiere," *Variety*, April 29, 2015, accessed January 20, 2017, http://variety.com/2015/tv/news/nbc-embraces-binge-viewing-releasing-all-aquarius-episodes-online-after-premiere-1201484383/.

5. "Cutting the Cord," *The Economist*, July 16, 2016, accessed January 20, 2017, http://www.economist.com/news/business/21702177-television-last-having-its-digital-revolution-moment-cutting-cord.

6. For more on the progress of cable's attempts to combat Netflix see Daniel B. Kline, "Can Netflix Save Cable?" *The Motley Fool*, November 7, 2016, accessed January 20, 2017, https://www.fool.com/investing/2016/11/07/can-netflix-save-cable.aspx and Kenneth Ziffren, "How TV Can Weather the 'Skinny Bundle' Storm of Streaming Services," *The Hollywood Reporter*, October 13, 2016, accessed January 20, 2017, http://www.hollywoodreporter.com/news/how-tv-can-weather-skinny-937142.

7. Derek Kompare, "Past Media, Present Flows," *Flow* 21 (2014), https://www.flowjournal.org/2014/09/past-media-present-flows/.

8. Edward Wyatt and Noam Cohen, "Comcast and Netflix Reach Deal on Service," *New York Times*, February 23, 2014, accessed January 20, 2017, https://www.nytimes.com/2014/02/24/business/media/comcast-and-netflix-reach-a-streaming-agreement.html?_r=0.

9. Julia Alexander, "Netflix's Chief Has Every Right to Be Worried About Net Neutrality in a Trump Administration," *Polygon*, December 1, 2016, accessed January 20, 2017, http://www.polygon.com/2016/12/1/13806052/netflix-streaming-net-neutrality-att.

10. Julia Greenberg, "Netflix's VPN Ban Isn't Good for Anyone—Especially Netflix," *Wired*, January 16, 2016, accessed January 20, 2017, https://www.wired.com/2016/01/netflixs-vpn-ban-isnt-good-for-anyone-especially-netflix/; Evan Elkins, "The United States of America: Geoblocking in a Privileged Market," in *Geoblocking and Global Video Cultures*, ed. Ramon Lobato and James Meese (Amsterdam: Institute of Networked Cultures, 2016), 190–199.

11. Mike McPhate, "California Today: Fretting Over the 'Netflix Tax,'" *New York Times*, November 28, 2016, accessed January 20, 2017, https://www.nytimes.com/2016/11/28/us/california-today-netflix-tax-video-streaming.html; Roberto Baldwin, "Chicago Kicks in 'Cloud Tax' on Streaming Services Like Netflix," *Engadget*, July 2, 2015, accessed January 20, 2017, https://www.engadget.com/2015/07/02/chicago-netflix-tax/.

12. Lisa Gitelman, *Always Already New: Media, History, and the Data of Culture* (Cambridge: MIT Press, 2006), 5.

13. *Ibid.*, 7.

From Primetime to Anytime

Streaming Video, Temporality and the Future of Communal Television

JUSTIN GRANDINETTI

When television critic Andy Greenwald of the sports and pop culture website *Grantland* reviewed the first season of the Netflix original series *House of Cards* (2013–) one month after its release, he remarked, "I finished watching the 13th and final episode of the first season on Sunday night and I have no idea if that makes me ahead of the curve or hopelessly behind."[1] This comment represents the recent conundrum of many entertainment critics, reporters, and bloggers, in that the temporal regularity of television has been thoroughly interrupted by the new Netflix model of releasing an entire season at a time. More importantly, this release model also affects viewers, reshaping the communal television experience amid the digital streaming revolution.

In February 2012, Netflix began its foray into original programming with the Norwegian series *Lilyhammer* (2012–2014). *House of Cards*, a political drama starring Kevin Spacey and Robin Wright, debuted a year later. Contrary to the weekly scheduling of linear television, all 13 *House of Cards* episodes were available to viewers the day of release. Netflix has since distributed a growing number of series in this manner, including the fourth season of the cult classic *Arrested Development* (2003–2006, 2013) and the popular original series *Orange Is the New Black* (2013–; referenced onward as *OITNB*). Although these series have garnered critical acclaim, they simultaneously pose new questions to the continued existence of communal television viewing. Most notably, the Netflix release model creates questions of how both viewers and news organizations adjust to this new reality as well as the ways the reactions of these groups consequently shape future Netflix programming.

In the following essay, I rhetorically analyze articles by publications such

11

as *Vulture*, *The A.V. Club*, and *Grantland* to argue that the binge-watching behavior encouraged by streaming platforms like Netflix fundamentally reshapes the nature of communal television viewing and introduces an antagonism that was absent in earlier iterations of "Web 2.0" audience engagement. More specifically, the interactions between the entertainment correspondents and viewers on these websites provides a space to extend theories of rhetorical vernacular posited by Gerard Hauser, Erin McClellan, and Ralph Cintron, as these online engagements represent new developments of changing temporality and communal experience of television.

Of course, these events do not operate in a purely cause and effect manner. Audience response to Netflix's all-at-once release strategy will continue to impact future production and distribution strategies, and ultimately guide the discourse among standing and burgeoning online communities. As such, this essay contributes to the interdisciplinary conversations about how interactive electronic environments impact the texture of publics. Furthermore, as *Arrested Development* and *OITNB* were among first Netflix original series releases, the reactions analyzed in this essay represent some of the vanguard responses to the disruption of television temporality. Both the critical and viewer response to these series contributed their online ubiquity, and this pervasive dialectic sets these programs apart from the few vanguard Netflix releases that preceded them.

In order to understand audience reactions to this changing temporality of the television experience, it is important to examine theories of the way rhetoric functions in these quotidian spaces. As such, this essay focuses on user commentary in public and communal forums—the most encompassing of which is the Internet. In his work contemporary ethnography, Cintron writes, "The availability of critical inquiry to everyone is a position that is important to take. Quite simply, rhetorical analysis is not just an endeavor of specialists but also an everyday affair."[2] In these colloquial and communal spaces of the Internet comment sections, audiences of Netflix programming create meaning and negotiate this changing television temporality together. Furthermore, the evolving vernacular rhetoric of these spaces follows Hauser's theories, in that

> ordinary people, whether they are neighbors or a class, develop and rely on a vernacular with rhetorical salience. Although not a discourse of power and officialdom, it nonetheless adheres to the fundamental rhetorical demand for propriety ... vernacular exchanges indicate bonds of affiliation; they speak a legible and intelligible rhetoric of shared values and solidarity.[3]

Before continuing, it is also important to briefly define the term "audiences." Broadly speaking, the audience for these programs is any individual who chooses to watch them. However, when discussing the communal discussion that surrounds Netflix programming, there is a difference between viewers who watch the series at their leisure and entertainment correspondents who write about the programming professionally. This demarcation is critical in this essay, as the power relationship between these two audiences is often unequal. Nevertheless, in the publications I selected for examination, it is clear that all audiences of Netflix programming have the opportunity to contribute to the evolving nature of the communal television experience and communicate with one another using a negotiated vernacular. Furthermore, these

> studies of vernacular rhetoric, in all its forms, have the potential to expose sense-making processes in more raw and revealing ways than its official or formal rhetorical counterparts considered alone. It takes rhetorical criticism in an empirical direction, which includes not only audience judgments of "official" rhetoric but the construction of a rhetoric culture of their own.[4]

Taken together, investigation into the rhetorical strategies of the users of these online discourses provides a method of understanding the evolution of the communal television experience in a binge-watch paradigm fostered by online platforms such as Netflix.

Evolving Vernacular Rhetoric and the Internet

In the remainder of this essay, I investigate specific examples of the evolving communal television experience through an examination of three publications, *Vulture*, *The A.V. Club*, and *Grantland*. These websites were chosen as they represent the genre of entertainment reviews delivered in electronic environments. Moreover, in each of these spaces, writers and/or critics openly reflect about their decision to impose an order to the chaos created by attempting to review a television series released in its entirety. Both *Vulture* and *The A.V. Club* encourage audiences to comment on the site's reviews, which allows for an inquiry into the evolving communal television experience. The more traditional television release schedule that has dominated the majority of the medium's existence involves serialized programming. In this familiar paradigm, television series premiere new episodes in what is typically a weekly release schedule. Successful programs run for multiple seasons, in which a number of new episodes are released once a week across a number of months. However, this model's longevity appears to be in question, as new data points to the preference of audiences to consume content in so-called "binge" sessions.[5]

Television producers and showrunners have begun to recognize the preferred methods of audience consumption, and as such, now model their series to further promote this. A 2013 issue of *Newsweek* cites conversations with "the people behind *Breaking Bad* [2008–2013], *Game of Thrones* [2011–], and so on, and it soon becomes clear that they've designed these series to be more bingeable—more propulsive and page-turning—than anything the networks ever pushed on us in the past."[6] Additionally, *Breaking Bad* creator Vince Gilligan explains an embrace of binge-watching: "I've always said that I don't see my show as serialized so much as hyperserialized. That is something that, honestly, I wouldn't have been allowed to do 10 or 15 years ago."[7] Consequently, binge-watching can be viewed as a collaborative negotiation between multiple entities—audiences, creators, and streaming platforms. Furthermore, studies of temporal displacement for audiences suggests that this "simultaneity within the structure of [many television programs] … parallels a growing concern of audiences with the immateriality and timelessness of interactive online digital technologies."[8] As such, marathon consumption of television programming is a natural evolution of the television experience, propelled forward by an increased rhetorical awareness of audiences by producers and amplified availability of programming via streaming platforms.

Though this television streaming revolution occurs in concert with human nature and desire, it is not without conflicting consequences. Viewers are caught between the contemporary thirst to binge-watch programming and the long-established desire to be part of a large community through television. This conflict has given rise to a myriad of negotiations between audiences and the traditional television model, social media, and blogs. Simply put, the polar opposites of the desire for the community that the traditional television experience provides and the human desire—likely hastened by a rapidly changing technological landscape—to consume content in binge sessions has led to crossroads for the future of television. Subsequently, traditional television, audiences, and steaming platforms will in turn continue to adapt to changing notions of temporality.

These spaces of inquiry follow Cintron's contention that "rhetorical analysis can help make sense of everyday language use … rhetorical analysis need not be about famous beaches and or the written word. Indeed, it need not be about the discursive at all and should also include the non-discursive and performative."[9] As such, audience and blogger commentary are ideal areas of study for understanding the negotiation of the new temporality and vernacular rhetoric created in response to the binge-watching paradigm created by the release of Netflix original series. When conducting my analysis on the responses to *Arrested Development* and *OITNB*, four themes quickly emerged: negotiation of temporality, responses to binge-watching, navigating spoilers, and cultivation of online ethos.

Negotiation of Temporality

Research into the temporality of television has investigated the ways in which awareness and anticipation of the viewing habits of others impacts the coordination and production of the collective television experience.[10] In addition, theorists have pointed to an increased use of disrupted temporality in television narratives. For example, the narrative of season four of *Arrested Development* is told through episodes that focus on the program's specific characters. Individual episodes do not constitute a linear story—instead, viewers are asked to consume the series holistically in order to fully understand the overarching storyline (and many of the recurring jokes). It is possible that this temporal displacement is a result of postmodern schizophrenia, in which "there is no sense of the 'past' or 'future' but rather an instantaneous and vacuous sense of the 'present.'"[11] This "vacuous present" comes as a result of the overwhelming availability of information and choice in an individual's free time. Consequently, the temporal misalignment depicted on many contemporary programs increasingly mirrors the disjointed nature of the communal television experience. Despite this apparent acceptance of a loss of temporal frame of reference, audiences continue to adapt to synchronize consumptive viewing patterns.

Studies have found that new technologies influence viewing patterns of American households in that viewing follows certain rhythms.[12] Moreover, audiences use technologies in order to fill time between viewing activities, but they can also make this time more or less meaningful through the use of on-demand technologies.[13] In addition, viewing patterns are "are influenced by larger social groups—groups that extend beyond a single household."[14] While it would follow that increased flexibility of television viewing through the use of handheld devices and high-speed internet allows for more personalized schedule, it also appears that audiences remain keenly aware of the viewing habits of others.

In its coverage of season one of *OITNB*, *The A.V. Club* examined two episodes at a time. These two-episode reviews were released every seven days, which created an imposed schedule on those who wished to discuss the series in the comment sections. However, an analysis of the comment sections quickly demonstrates the challenges in attempting to impose this two-episode limitation on viewers, many of whom had binge-watched the entire season. When responding to the comments of others, individuals that were further along in their viewing had trouble keeping track of what had occurred in the first two episodes of a 13-episode season. In one such interaction, user K. Thrace had to question when a recurring character appeared, writing, "Was the nun even in that episode? I forget. She's great, anyway." Other comments mirrored this sentiment, in that while viewers who binge-watched the series seemed to desire to discuss *OITNB* in a holistic sense, *The A.V. Club*'s review structure imposed a more traditional paradigm on viewers.

Commenters occasionally expressed other issues with the site's insistence of reviewing only two episodes a week, as evidenced by the following interaction between critic Myles McNutt and a site commenter:

> **MylesMcNutt:** The downside of covering the first two episodes is that it encourages you to stop after two episodes when I'd argue the third episode is the strongest—or at least most distinctive—of the ones I've watched. I'd encourage you to check it out, and see if it changes your outlook.
>
> **mouse clicker:** Well, I had already watched the first two episodes and kind of decided not to keep going before I read the review. My wife really seems to like it, so I'll probably give the third episode a shot, too. But she was also hooked by *Weeds* [2005–2012] immediately and I didn't start actually enjoying that show at all until the second season.

Here, it is evident that commenters had issues with the method of review employed by *The A.V. Club*. Those who felt the series blossomed in episode three or later expressed similar complaints with the site not taking a complete approach in order to encourage viewers to keep watching. These interactions create "an agonistic zone between official and mundane communication in which the established and the marginalized vie for power. Their struggle is enacted through contrasting rhetorical modalities seeking public allegiance and legitimation."[15] Thus, while *The A.V. Club* uses its position to enact a more traditional viewing paradigm, some viewers took it upon themselves to advocate for the series in a more complete manner, attempting to persuade others to continue watching (and even to catch up through binge-watching).

On the subject of *OITNB*, writers for *Grantland* had high praise for the series, but varying affection toward Netflix's release model. Some of the site's contributors lauded the binge-watch model, with Rembert Browne arguing, "Netflix banked on what its entire existence was built on, binge-watching, and came out looking like a genius. The company's easily one of the comeback stories of the year."[16] However, others expressed displeasure toward various aspects of the distribution of programs such as *OITNB*. In a look back at the year in television, Greenwald wrote, "*Orange* restored my faith in TV's ability to tell diverse stories in exciting new ways, and I wanted to shout as much from the rooftops, or at least the guard tower. Too bad I waited until late in the fall to finish the season. By then, there was no one left to tell."[17] Ultimately, Greenwald and others have complaints not about *OITNB* itself, but instead about the Netflix release model. In his preview of the series from earlier in 2013, Greenwald noted,

> I hate the Netflix distribution model, and the way, every few weeks, it backs up the content truck at the stroke of midnight and offloads 13 hours of industrial-strength television. I've written at length about how this emphasis on quantity robs quality

material of the time necessary to consider or even savor it, how it mutes the great beehive of conversation that has sprung up around TV in recent years and replaced it with the lonely, furtive clicks of a solitary remote control. We don't always eat for the sole purpose of getting full, and we shouldn't consume art that way, either. There's a reason you'd never order anything à la carte that can be found in an all-you-can-eat buffet.[18]

Though disrupted temporality is seemingly embraced by fans, who—despite some complaints—rabidly consume Netflix series in marathons, critics and bloggers are less enthusiastic. In these instances, blogs/websites represent what Hauser calls a site of research "where the vernacular rhetoric of place and space is performed, such as a city square where the local culture of inclusion and exclusion, of use and abuse of community spaces is enacted on a daily basis, and where those who are using the city square can say for themselves what they think they are doing by their public performances."[19] In the traditional television review model, the power and control of space is determined by writers, who are in turn supported by the temporal confines established by networks and channels. However, the binge-watching model allows greater power for viewers, who are likely to watch series on their own schedules—regardless of the impositions of the entertainment publishing world.

Writers under the traditional television model are in a position of power over regular viewers. Often, those writing about television are privy to the content early. As such, they are afforded a chance to carefully formulate their own opinions about the program before the majority of viewers have seen it. This, in turn, works to as a way to cultivate professional credibility. In the case of new programming, early reviews have a chance to alter the perception of readers, who are then open to the possibility that their own experience with a program has been shaped by the opinions of others. Under the Netflix model, entertainment correspondents are given the same release as viewers (or at least only given access to a few episodes a day or two before the full season is released). Therefore, it is conceivable that some of the viewers who are responding in the comment sections are more knowledgeable—and further into the season—than the authors of the articles. While the possibility of a reversal of roles (and by extension, authority) between critics and readers may seem subtle and benign, this impact is nevertheless a developing externality of the evolution of the communal television experience.

Responses to Binge-Watching

Though sometimes viewed as a solidary and anti-social activity, watching television is, in actuality, a largely communal event. Television has "traditionally been an important facilitator for social interaction and a popular source of conversation."[20] Moreover, the advent of social networking sites allows audiences to transcend physical barriers and watch television "together." This new practice of watching television combines two elements, viewing and the real-time sharing of reactions and responses via the Internet, resulting in a pseudo-communal viewing experience.[21] This negotiation of temporality is not a new aspect of television. The medium has always united audiences across space; however, technological advances disrupt the fixed temporal notion of time.

Numerous scholars have discussed the concept of "liveness" in relation to the existence and development of television and the communal television experience. Nick Couldry notes that liveness "is a category whose use naturalizes the general idea that, through the media, we achieve a shared attention to the 'realities' that matter for us as a society."[22] As such, liveness via live transmission connects viewers to shared social realities as they are occurring and helps to create realities that matter to viewers.[23] Nevertheless, changes in television technologies have shifted the formulation of liveness. For instance, Graeme Turner postulates that while many deem liveness and sharedness essential to the experience of television, this attribute has been interrupted since the widespread ability for viewers to record and replay programs.[24] Meanwhile, Chuck Tryon notes that these new technological practices can upset the social ritual of watching television; however, the "water cooler" status of television discussion persists largely due to social media tools such as Twitter and so-called "check-in" services such as GetGlue that encourage live viewing.[25] Taken together, there is a general agreement that

> liveness, in its most general sense of continuous connectedness, is hardly likely to disappear as a prized feature of contemporary media, because it is a category closely linked to media's role in the temporal and spatial organization of the social world. The category "liveness" helps to shape the disposition to remain "connected" in all its forms, even though (as we have seen) the types of liveness are now pulling in different directions.[26]

Liveness is essential to the television and the communal television experience, but these newer, altered forms of liveness continue to be driven by significant technological change, including the high-speed streaming technology of Netflix.

Netflix programming, free from the confines of a weekly schedule, is the antithesis of the traditional sense of liveness and communal television. Not all audience members have responded to this novel form of distribution, however. Netflix's model may represent the future of television. It may also lack the essential temporal regularity and subsequent sharedness that viewers have come to expect from their television experience. We are, in short, at a crossroads of sorts in terms of the social dynamics of television watching.

The concept of binge-watching is critical to the streaming model of television consumption. A 2013 Harris Interactive study found that "nearly 80 percent of US adults with Internet access watch TV through subscription on-demand services (like Netflix or Hulu), through cable on demand, or through a time-shifting device like a DVR. Sixty-two percent of people who watch TV whenever they feel like it will watch multiple episodes back to back."[27] Moreover, some argue that binge-watching is a kind of symptom of human evolution and anatomy. Richard Rosenthal, chairman of psychiatry at St. Luke's–Roosevelt Hospital Center in New York, explains that "whether you're deciding to watch 'just one more' episode of *Breaking Bad* or you're throwing back 'just one more' tequila shot, a similar sequence is playing out in that particular part of the brain."[28] The release of dopamine in the brain, combined with easier access to full seasons of television programming is considered one explanation for the rise of binge-watching. Taken together, this data and physiological reasoning points to a continued growth in both the use of streaming services and this indulgent method of consumption.

Analysis of the entertainment site *Vulture* demonstrates the ways that audiences negotiate the unfamiliar binge-watching format. Like *The A.V. Club*, *Vulture* reviewers imposed a semi-regular release of their individual episode reviews of *Orange Is the New Black*, each one coming within three to five days of the previous episode recap. Commenters on this site were more open in voicing their feelings toward Netflix's distribution strategy. The following selections are taken from the site's first episode review.

> **Jarira:** God damn this f*cking show for ruining my life. I had the best/worst bingefest this weekend. My eyes are burning but it was worth it.
>
> **typicaliowa:** You recapping this show just made my week. I finished binge-watching the season on Monday and have been in withdrawal since then. I can't wait to relive/dissect every episode!
>
> **Bookles:** I thought the whole series was very well done. It was so well executed that I finally discovered a problem with Netflix's all at once format. Now I have to wait until who knows when to see the next season. At least if it was traditional format the new season might already be filming. I guess I'll have to re-watch until then.

These comments demonstrate the conflicted relationship viewers have with *OITNB*, or at least its distribution. While the audience seems to lament binge-watching, it is often in jest. The real complaint is generally that the series was so addictive that individuals felt compelled to continue watching, usually at the expense of their free time. The final comment is telling in that although the viewer liked the series, s/he feels that the traditional temporal patterns of a waiting for a new season had been heavily interrupted by the binge model.

Additionally, viewers had to create informal communal rules for spoilers and posting etiquette. These interactions, however, did not always go without conflict, as evidenced in these exchanges between two commenters, followed by a response from editor Gilbert Cruz:

> **misspam:** Watch out for the spoilers. Not everybody's seen the first episode. Are recaps really the thing for a watch-one-or-binge format?
>
> **im10ashus:** @misspam43—So, they should wait for everyone to catch up before recapping? You do realize they recap a lot of shows here, right?
>
> **Gilbert Cruz:** If you hadn't seen the first episode, you wouldn't have clicked on the article, though, right?

Here, user misspam attempts to warn the community that many of the posts in the comment section contain spoilers. misspam also muses whether or not the site's reviews are done in the best possible manner for a binge series. However, misspam's comment is met with both sarcastic and pragmatic responses—both of which demonstrate the opinion that it is the responsibility of the individual to either keep up (or ahead) of *Vulture*'s reviews, or to simply not read the article. As such, it is clear that *Vulture*'s reviews represent a continually evolving space of rhetorical vernacular. In this space, both the writer and users express their opinions as to the best practices when reviewing and discussing a series free from traditional television temporality. Consequently, the variety of responses signifies the unsettled nature of this vernacular space. The consequences of the site's decision to impose a temporality on viewers are a point of tension, and negotiated within the comment section. Moreover, these comments also engage with the discussion of whether or not the onus is on the individual, the community, or the publication itself to avoid spoiling details for others.

Finally, some commenters exhibit frustration with either other posters or the rhetorical vernacular confines of the comment sections. In one such comment, user dgoings writes, "Nobody cares if you binged watched. If it doesn't have to do with the episode that is being RECAPPED keep it to yourselves." While this comment demonstrates anger toward other users who are commenting on episodes beyond the ones being reviewed, many other individuals are undeterred. A large portion of commenters expressed excitement in discussing future episodes of the series, despite the fact that the particular review was read by users who had viewed highly varying numbers of episodes.

Additionally, the comment sections of *Vulture* further demonstrate user negotiation of the disrupted temporality of the communal television experience. These continuous commenter interactions form what Hauser observes as "the dramaturgy of vernacular exchange offers deep insight into the rhetorical performance of a movement that is missed on studies focused exclusively on the formal rhetoric of leaders."[29] In the rhetorical communities of these comment sections, users must navigate emerging rules of acceptable vernacular and dialogue. There are attempts by the community (via both *Vulture*'s review structure and user comments) to create order out of the temporal chaos inspired by Netflix's all-at-once distribution strategy, but individual commenters nevertheless fail to conform to these expectations. While some are apologetic in their transgressions, others openly question the wisdom of reviewing *OITNB* in the same methodical manner as regularly scheduled television.

An examination of *Grantland* gives insight into the mindset of writers and editors who are forced to recalibrate their coverage of Netflix programming. For media criticism, this new distribution strategy represents "radically new ways of thinking about human symbolic activity moved beyond the podium to the streets where the micro-practices of moment-by-moment interactions contribute not only to the organic character of the culture but become a significant source of rhetorically salient meaning and influence."[30] In this way, writers must try to reestablish a regular temporality to streaming content distribution, a medium that eschews this—or any—kind of structure.

Grantland addressed Netflix's *Arrested Development* new season by reviewing two episodes each week. Despite the potential for binge-watching, the site imposed a weekly schedule that mostly approximates the typical once-a-week episodic distribution schedule. In the review of the first episode of season four of *Arrested Development*, Ana Marie Cox writes, "Did you binge all weekend on the new *Arrested Development* episodes? To each his own. We're going to slow it down a little. Two episodes per week, one at a time. It's what Mitch would want."[31] Here, "Mitch" refers to *Arrested Development* creator Mitch Hurwitz, who before the Netflix release, stated that

> I'm really doing everything I can to put out that misconception that it can be watched in any order. Although I really did have that ambition at one point.... Not only will the episodes be available at the same time on Netflix, but they also cover the same period of time in the characters' lives. So it seemed like, yeah, you should be able to jump in in any order and see George Michael's episode and then maybe Buster's episode, if you want [but] I pretty quickly realized everything here is about the order of telling the stories, that there will be shows where you find out a little bit of information and then later shows where you revisit the scene and you find out more information.... I thought, okay, this may not be up for debate.[32]

Hurwitz also responded to a question about binge-watching the series, stating, "No one should binge-watch" because "you'll get tired." With the creator's firm perspective on how viewers should experience his series in mind, *Grantland* imposed a semi-regular schedule on its readership. Though viewers are able to feast at will on series, it could, potentially, come at the cost of the more traditional communal aspects of television viewing. The idea of a community of viewers watching simultaneously has an innate and engrained power for American viewers, a power that Hurwitz and *Grantland* urged viewers to remember when watching *Arrested Development*. (It is also worth noting here that publications like *Grantland* prefer the normal distribu-tion model because it enables them to publish individual episodic reviews that extend the window for web traffic and clicks across a longer period of time.)

Navigating Spoilers

Viewers who fail to watch television programs quickly after their air date or release are opening themselves to the possibility that major events will be prematurely revealed, or "spoiled" for them as they peruse social media. In the comment section of *The A.V. Club*, users quickly adopted the vernacular rhetoric of spoiler etiquette. For most users, this community con-sideration involved beginning their comment with some use of the word "SPOILERS" in all capital letters. This acts as an unmistakable warning sign for others who are not as far along in the series, and encourages them not to read the post, lest they incur the risk of having a major plot point revealed. When this etiquette was not followed, users were quick to voice displeasure at these vernacular transgressions.

> YO MOMA...cuz im secure like t: Ok, where is everyone. I was going to start into episode 10, but I'm afraid my parents will hear the overtly sexual content, so ... it'll have to wait for tomorrow.
> I'm really, really loving it. I think I could partly attribute this to the "marathon high" though. I can't really remember what happens in any one episode.
> EDIT: Really sorry for the spoilers. The review for the first six seemed pretty dormant. Again, sorry.
> MylesMcNutt: I hate to be that guy, but I'm going to—based on the fact I saw your spoiler and wished I hadn't—suggest perhaps editing that out. I'd rather not have a thread discussing future plot details in any specific detail, especially not anything tied to major character developments.
> I get you want to talk about where you are. I wish I could say that these reviews could move at the same pace. But for now, the full season review might need to be the spot for that, if you can be a bit patient.

In this interaction, user YO MOMA is publically chided for not following the developing rules of conversation in the site's comment section. As demon-strated by the edited comment, s/he apologizes for this transgression multiple times, and in an attempt to make amends, has removed the egregious spoiler from the comment. This interaction exemplifies Hauser and McClellan's notion that "we enact vernacular performances that either uphold the status quo or blatantly disregard it. Such performances necessarily impact larger understandings of how to interact with strangers."[33] In this virtual community the rules of vernacular interaction are upheld much in the same way they would be in face-to-face conversation. Though this is just one selected example of such an interaction, similar online exchanges occur regularly.

Finally, users employed either the aforementioned spoiler notifications or some form thereof to cultivate ethos, or credibility. Many began their post with a line such as "I am only three episodes in" or "Just finished the whole series" in order to mark their post as ahead of *The A.V. Club*'s review structure. Ultimately, this strategy was replicated in the future reviews of *OITNB* on *The A.V. Club*'s website, as it serves as context for readers. Hauser postulates that these "pragmatic functions, such as temporal references, could indicate a time not captured by the literal meaning of the words but by their context."[34] As commenters exercise their desire to become part of this new communal television discussion, it is critical that they adopt the vernacular of their discourse community and ensure their comments contain an awareness of the context of both spoilers and the differing viewing patterns of others.

Cultivation of Online Ethos

In digital spaces, the credibility of an individual is often impossible to discern. While writers have the ability to reveal their identity (and by extension their credentials), users who post to comment sections are not afforded this rhetorical appeal. However, in the examined websites, users were able to quickly develop strategies to cultivate ethos in order to persuade fellow commenters. An investigation of the comment sections of *The A.V. Club*'s reviews of season four of *Arrested Development* further understanding of the way that these interactions are spaces for the creation of rhetoric. Hauser's vernacular rhetoric "widens the scope of rhetoric to include instances of vernacular exchange, direct attention to collective reasoning processes as they are disclosed in vernacular exchanges, locate public opinion and processes of creating common understanding, and regards the dialogue of vernacular talk as a significant way by which public opinion is developed."[35] The creation of vernacular rhetoric is apparent in the following selected comments and exchanges from *The A.V. Club*.

Many comments on *Arrested Development* reviews focus on viewer disappointment when watching the early episodes of the series. Responding to concerns of viewers who had only watched the first few episodes, commenters on the site's first *Arrested Development* review worked to influence the opinions of others and convince them to give the series another chance, as exemplified by the following exchange:

> **Tim Lieder:** I thought so too. It was also a big part of why I didn't like the first episode—because even though I agree with the writers of this column that

> Michael is just as crazy and as self-centered as the rest of the Bluth clan, I
> think it's much funnier when his insanity is just below the surface and he has
> that veneer of straight man.
> Then I saw all the episodes and started rewatching the show. Now it's hilarious.
> **ButlerWhoGooglesThings:** Knowing where everything is going makes these two
> episodes much better. I'm not saying they were terrible as stand-alone
> episodes, but just that they are improved with context.
> George Sr. in the "sweat and squeeze" con was a perfect character moment
> and Michael getting voted out was as strong a scene as anything they've ever
> done.

These comments exemplify Hauser's assertion that vernacular rhetoric helps sway public opinion. Commenters who want others to continue watching the series first commiserate through an initial and shared disappointment. They then encourage their peers to continue watching by stating their pleasure with the progression of the season, often by foreshadowing some of the particularly appealing humorous situations.

Additionally, through mention of their progression through the *Arrested Development* season, commenters are able to cultivate comment section ethos:

> **Tearinitup Drifter:** Not sure if this helps, but the stuff at college is happening six
> months prior to the events in the opening scenes in episode one (I think…).
> I've only seen these two episodes.
> **spicoli323:** Having seen almost the whole season … you are correct about this.
> **MothaF_NDixon:** Having seen 10 episodes now, these earlier episodes certainly
> "age" well after you get more info and scenes that come later. I'll definitely be
> rewatching this again, as it appears it may be as enjoyable as the first watch but
> for completely different reasons.

In this interaction, the replies begin with users couching their statement with "Having seen." This notification creates a hierarchy of understanding and credibility when it comes to comments on *Arrested Development*. Moreover, the use of similar language demonstrates the way that vernacular rhetoric in this particular public space is quickly adopted by others, eager to give order to the chaos of the disrupted temporarily of the communal television experience.

An examination of the *Vulture* comment section demonstrates a continuation of themes found in the comments of *The A.V. Club*. However, unlike the reviews structure of *The A.V. Club*, *Vulture*'s reviewers released 15 reviews (one for each episode of season four of *Arrested Development*) all on the same day, with promises of more in-depth recaps in the coming weeks. Thus, the site was able to satisfy fans demanding immediate spaces of discourse without imposing a more traditional weekly or bi-weekly limitation.

In these spaces, commenters continued to display forms of encouragement for fellow viewers, supplemented through an ethos generated from their progression through the series. Interacting with critic and fellow commenter Matt Zoller Seitz, johnnyb0731 wrote,

> I'm glad that you enjoyed the season. I watched it all on Sunday and while I thought the first two episodes were slowish I loved the season after that. I think it took some adjusting to the difference in structures and how the jokes were going to be set up and layered together. I've been very surprised to see the negative reactions by a lot of the critics because I think that the show came back better than I could have dreamed it would.

Here, the user is able to establish their credibility by stating that s/he not only watched the entirety of season four, but also did so in a single day. Though the comment is directed at creating dialogue with a particular user, johnnyb0731 acknowledges widespread community concerns that the series "starts slowly," which helps encourage those hesitant to continue to keep watching. This encouragement continues with an interaction between two users in the site's review of the first episode:

> **AaronKT:** I didn't much care for this episode the first time around. It took me a few episodes to get into it. But having watched them all and rewatching episode 1, I giggled all the way through this episode. If anything this season is even more dense then it used to be…. Showstealer pro trail version was the best gag this episode.
>
> **backinstolaf:** @AaronKT Yes, once you watch the whole season and start rewatching you catch a lot of small things that make it so much better!!

Yet again, it is clear that many users desire a space to discuss the series in its entirety. However, interaction with one another is difficult, as there is potential for spoilers or lack of frame of reference, as some commenters may not have yet watched the entire series. As such, how many episodes an individual has watched becomes an important qualifier and self-identifier in this vernacular community.

Overall, these adaptations by both writers and commenters represent the rudimentary stage of the post-temporal television world. Nevertheless, it appears that adjustments by entertainment publications to Netflix and other streaming services will only continue. The unsettled nature of time and viewer participation creates a myriad of problems for writers and audiences alike, who now have no frame of reference as to the progress that others have made in their favorite programs. Taken together, "studies of vernacular rhetoric offer news about the ways in which ongoing social discourse serves as a mode of influence on what people think and do."[36] By examining the interaction between commenters and critics in these online discourse communities, it is possible to gain a better understanding of the formation of vernacular rhetoric. Ultimately, these examinations illuminate the ongoing negotiation of temporality and the communal television experience created by Netflix's release model.

Conclusion

This essay's examination of audience and critic responses to Netflix's distribution practices provides insight to the negotiated rhetorical vernacular of the new communal television viewing experience. The traditional communal television experience is predicated upon temporal confines created by a regular episode release schedule. With the advent of the Internet as a space for communal discussion, this audience regulation was reinforced, as audience members were able to take part in regularly scheduled discussions with more users than ever before. However, Netflix's practices have disrupted not only the temporality of the longstanding distribution model, but also the communal experience of discussion. While this essay examines some of these developing reactions, it should be noted that these sites, reviews, Netflix series, and comments are merely a fraction of the dialogue that continues to regularly occur online. Nevertheless, these interactions represent new developments regarding the changing temporality of communal television and the concomitant communal television experience.

When taken in the context of the history of television, Netflix's all-at-once distribution strategy is still in its nascent stages. It seems clear the Netflix as a brand and content producer understands audience in ways never before realized. Networks under traditional television model are at an inherent disadvantage compared to online streaming services that have access to enormous amounts of consumer data. Conversely, Netflix has been able to leverage this advantage in order to grow in popularity and release near guaranteed hit programs. Traditional television, meanwhile, continues a long-established haphazard model for new releases, while Netflix is able to gather data about the programs people are watching in new and efficient ways. Recent data claims that 35 percent of all U.S. consumers ages 13 to 54 state that they use Netflix at least once a month.[37] Subscribers watch an average of five series and four movies a week.[38] In essence, "Netflix doesn't know merely what we're watching, but when, where and with what kind of device we're watching. It keeps a record of every time we pause the action—or rewind, or fast-forward—and how many of us abandon a show entirely after watching for a few minutes."[39]

In many ways, then, the "risk" of spending $100 million to create *House of Cards* was not a risk at all. Netflix knew that there was an audience that loved the original BBC version as well as director David Fincher and actor Kevin Spacey. As such, there was no gamble; the algorithms demonstrated that an audience existed for this program. How Netflix continues to adapt to audience demands—the simultaneous and yet conflicting desires for both binge-able, all-at-one television releases and a communal television experience—remains to be seen. Nevertheless, many of the recent actions taken by Netflix demonstrate a critical awareness of the conflict by the media giant.

Netflix does appear to be aware of the relationship between websites that write about television and the viewers that watch and comment on it. When releasing the second season of *House of Cards,* some reviewers indicated that Netflix required them to adhere to a heavy-handed embargo agreement that prohibited reviews of the first four episodes before the series' release date.[40] According to Pam Brown, the agreement included the vague, yet ominous statement, "I understand and agree that any breach of these conditions will cause irreparable harm for which recovery of money damages alone would be inadequate."[41] Here Netflix expressed concern that the season's surprises could be spoiled for viewers. In a larger sense, Netflix's actions continued to dismantle not only the traditional communal television model, but also the entrenched power relationship between the television critic and viewers. It is also worth noting that at least one writer defied Netflix's wishes and released a review of the first episode early.[42] It appears then, at least for now, that those whose careers depend on early access—and the ability to drive public opinion—are most concerned about the way Netflix has altered the familiar communal television model.

It should be noted that Netflix benefits by disrupting the communal television experience. The company's programming has been extremely successful, and the publicity garnered by each subsequent release generates continued interest in Netflix original series. Moreover, the aforementioned studies demonstrate that television audiences prefer binge-watching, which, while contrary to the traditional communal television model, is still an example of the media giant understanding and catering to audiences in a way that network and cable television cannot. Despite its existence as an agent of change, Netflix has instituted features that help to create a sense of communal experience. Movies and series on the media platform are sorted by categories including the long-standing "Popular in Your Area" and the more recently tested feature "Trending Now." As of April 2015, "Trending Now" is still in its beta test, and the feature is not available to all users. According to Netflix's official statement, the company notes that the feature "allows us to not only personalize this row based on the context like time of day and day of week, but also react to sudden changes in collective interests of members, due to real-world events such as Oscars or Halloween."[43] Though Netflix's non-live, binge-watch model may appear to be anathema to live events and a larger television community, Netflix seems aware of the audience's desire for greater connectivity and open to features that accomplish this feat.

Finally, it is worth noting that Netflix's influential distribution strategies have now begun to influence film releases as well. In a 2013 speech, Netflix's Ted Sarandos remarked, "The model that we're doing for TV should work for movies. Why not premiere movies the same day on Netflix that they are opening in theaters?"[44] This development is a monumental shift for how films are distributed, and will therefore surely have a myriad of impacts on the communal experience of movies. Though this change is beyond the scope of this research, it does represent a future area of study for scholars, as both television and movies are enveloped in the larger umbrella of entertainment. Moreover, it demonstrates that the impact of streaming services on the temporal nature of the communal experience of entertainment media is malleable and in constant flux due to rapid technological advances and desire for convenience.

The evolving temporal nature of the communal television experience represents a space to extend theories of rhetorical vernacular posited by Hauser, McClellan, and Cintron. As evidenced by the interactions on *Vulture*, *The A.V. Club*, and *Grantland*, the continuing interactions of critics, writers, and commenters represent the vanguard of a developing area for rhetorical examination. Furthermore, entertainment publications are still in unsettled territory in terms of negotiating an effective and acceptable communal space audience discussion.

Through a greater understanding of the audience, Netflix has been able to quickly rival the success of traditional television. Audiences engaging in binge-watching are lured by convenience, yet simultaneously forced to adapt their familiar communal experiences, perpetuated via social media, blogs, and other online mediums, in order to fit this new model. As such, Netflix has had a number of revolutionary impacts on temporality—both in the digital and physical world. Furthermore, the disruption of online communities has created a space for burgeoning rhetorical vernacular, as audience members and bloggers struggle to adapt to a loss of traditional regulation. While many of these changes are still in their early stages, it is clear that Netflix and other streaming media have permanently altered both temporality and the communal television experience.

NOTES

1. Andy Greenwald, "Isolated Power," *Grantland*, March 6, 2013, accessed November 3, 2014, http://grantland.com/features/netflix-house-cards-gamble/.
2. Ralph Cintron, "'Gates Locked' and the Violence of Fixation," in *Towards a Rhetoric of Everyday Life: New Directions in Research on Writing, Text, and Discourse*, ed. Martin Nystrand and John Duffy (Madison: University of Wisconsin Press, 2003), 14.
3. Gerard Hauser, "Attending the Vernacular. A Plea for an Ethnographical Rhetoric," in *The Rhetorical Emergence of Culture*, ed. Christian Meyer and Felix Girke (New York: Berghahn Books, 2011), 164.
4. Hauser, "Attending the Vernacular," 169.
5. Kelly West, "Unsurprising: Netflix Survey Indicates People Like to Binge-Watch TV," *Cinema Blend*, 2014, accessed October 27, 2015, http://www.cinemablend.com/television/Unsurprising-Netflix-Survey-Indicates-People-Like-Binge-Watch-TV-61045.html.
6. Andrew Romano, "Why You're Addicted to TV," *Newsweek*, May 15, 2013, accessed

November 3, 2014, http://www.newsweek.com/2013/05/15/why-youre-addicted-tv-237340.html.

7. *Ibid.*

8. Paul Booth, "Memories, Temporalities, Fictions: Temporal Displacement in Contemporary Television," *Television & New Media* 12.4 (2011): 373.

9. Cintron, "'Gates Locked' and the Violence of Fixation," 5.

10. Lilly Irani, Robin Jeffries, and Andrea Knight, "Rhythms and Plasticity: Television Temporality at Home," *Personal & Ubiquitous Computing* 14.7 (2010): 621.

11. Booth, "Memories, Temporalities, Fictions," 375.

12. Irani, Jeffries, and Knight, "Rhythms and Plasticity," 631.

13. *Ibid.*, 630.

14. *Ibid.*, 622.

15. Gerard Hauser and Erin Daina McClellan, "Vernacular Rhetoric and Social Movements: Performances and Resistance in the Rhetoric of the Everyday," in *Active Rhetoric: Composing A Rhetoric of Social Movements*, ed. Sharon Mackenzie Stevens and Patricia M. Malesh (Albany: State University of New York Press, 2009), 29.

16. Rembert Browne, "Who Won 2013?" *Grantland*, December 30, 2013, accessed November 3, 2014, http://grantland.com/features/rembert-browne-year-end-bracket/.

17. Greenwald, "The Year TV Got Small," *Grantland*, December 18, 2013, accessed November 3, 2014, http://grantland.com/features/breaking-bad-black-mirror-year-television/.

18. Greenwald, "The Great *Orange Is the New Black* Is Suddenly the Best Netflix Series Yet," *Grantland*, July 15, 2013, accessed November 3, 2014, http://grantland.com/hollywood-prospectus/the-great-orange-is-the-new-black-is-suddenly-the-best-netflix-binge-watch-series-yet/.

19. Hauser, "Attending the Vernacular," 169.

20. D. Yvette Wohn and Eun-Kyung Na, "Tweeting about TV: Sharing Television Viewing Experiences via Social Media Message Streams," *First Monday* 16.3–7 (2011), accessed December 9, 2014, http://www.firstmonday.org/ojs/index.php/fm/article/view/3368.

21. *Ibid.*

22. Nick Couldry, "Liveness, 'Reality,' and the Mediated Habitus from Television to the Mobile Phone," *Communication Review* 7 (2004): 355.

23. *Ibid.*, 355–356.

24. Graeme Turner, "'Liveness' and 'Sharedness' Outside the Box," *Flow* 13.11 (2011).

25. Chuck Tryon, *On-Demand Culture: Digital Delivery and the Future of Movies* (New Brunswick: Rutgers University Press, 2013).

26. Couldry, "Liveness, 'Reality,' and the Mediated Habitus," 360.

27. Jessica Goldstein, "Television Binge-Watching: If It Sounds so Bad Why Does It Feel so Good?" *The Washington Post*, June 6, 2013, accessed November 3, 2014, https://www.washingtonpost.com/lifestyle/style/television-binge-watching-if-it-sounds-so-bad-why-does-it-feel-so-good/2013/06/06/fd658ec0-c198-11e2-ab60-67bba7be7813_story.html.

28. *Ibid.*

29. Hauser and McClellan, "Vernacular Rhetoric and Social Movements," 26.

30. Hauser, "Attending the Vernacular," 159.

31. Ana Marie Cox, "*Arrested Development* Recap: A Slow-Binge on Season 4, Episode 1: 'Flight of the Phoenix,'" *Grantland*, May 28, 2013, accessed November 3, 2014, http://grantland.com/hollywood-prospectus/arrested-development-recap-a-slow-binge-on-season-4-episode-1-flight-of-the-phoenix/.

32. Amos Barshad, "Mitch Hurwitz PSA: Talk to Your Kids About Watching *Arrested Development* in Order, and One at a Time," *Grantland*, May 22, 2013, accessed November 3, 2014, http://grantland.com/hollywood-prospectus/mitch-hurwitz-psa-talk-to-your-kids-about-watching-arrested-development-in-order-and-one-at-a-time/.

33. Hauser and McClellan, "Vernacular Rhetoric and Social Movements," 33.

34. Hauser, "Attending the Vernacular," 157.

35. Hauser and McClellan, "Vernacular Rhetoric and Social Movements," 29.

36. Hauser, "Attending the Vernacular," 169.

37. "Netflix Has Changed Viewing Habits," *Advanced Television*, September 30, 2011, accessed November 3, 2014, http://advanced-television.com/2011/09/30/netflix-has-changed-viewing-habits/.

38. *Ibid.*

39. Andrew Leonard, "How Netflix Is Turning Viewers into Puppets," *Salon*, February 1, 2013, accessed November 3, 2014, http://www.salon.com/2013/02/01/how_netflix_is_turning_viewers_into_puppets/.

40. Brian Lowry, "TV Review: 'House of Cards'—Season Two," *Variety*, January 30, 2014, accessed May 28, 2015, http://variety.com/2014/digital/reviews/review-house-of-cards-1201076822/.

41. Pam Brown, "So Ruthless He Gets It Done," *The West Australian*, February 10, 2014, accessed May 28, 2015, https://au.news.yahoo.com/thewest/entertainment/a/21383587/so-ruthless-he-gets-it-done/.

42. Willa Paskin, "House of Cards' Second Season Is Even More Ridiculous Than the Last One. Thank Goodness for That," *Slate*, February 13, 2014, accessed May 28, 2015, http://www.slate.com/articles/arts/television/2014/02/house_of_cards_season_2_reviewed_netflix_s_prestige_drama_embraces_its_own.html.

43. Olivia Armstrong, "Netflix Introduces New "Trending Now" Suggestions," *Decider*, April 14, 2015, accessed May 28, 2015, http://decider.com/2015/04/14/netflix-trending-now-suggestions/.

44. Dominic Patten, "Netflix In Talks For Third Season Of 'House of Cards,'" *Deadline*, October 26, 2013, accessed May 1, 2015, http://deadline.com/2013/10/netflix–in-talks-for-third-season-of-house-of-cards-620952/.

Terms of Excess
Binge-Viewing as Epic-Viewing in the Netflix Era

DJOYMI BAKER

The term binge-viewing has been adopted by the television industry, popular press, and scholars to describe watching several consecutive episodes or even seasons of a program in one sitting.[1] Currently, 70 percent of U.S. viewers identify themselves as television bingers.[2] The term binge has been under-theorized in current scholarship, and its negative associations of excessive consumption warrant closer examination. This essay argues that Netflix's narrative, promotional, and release strategies are predicated on a spectator endurance that might more fruitfully be conceived of as epic-viewing. Netflix creates its brand profile by encouraging viewers to perceive its products as best experienced as one epic text, from its back catalog of licensed programs across a multitude of genres, to its in-house productions such as *House of Cards* (2013–), *Arrested Development* (2003–2006, 2013), or *Marco Polo* (2014–2016). Netflix positions itself as the bigger, better option, offering it all—and right now. When Netflix launched *House of Cards* on its streaming service in February 2013, it made the entire first season available at once. At the time, other providers offered new programs in installments, and only their old series could be accessed as complete seasons. Netflix delivered an immediate and continuous epic, creating a new paradigm.

Binge-Viewing

In 2014, OxfordDictionaries.com added a separate entry for "binge-watch" (including under this umbrella its close cousin "binge-view"), tracing its use as a verb to the 1990s, but noting in its press release:

Use of the word binge-watch has shown a steady increase over the past two years, with notable spikes in usage recorded around the Netflix releases of *House of Cards*, Season Two in February 2014 and *Orange Is the New Black*, Season Two in June 2014. According to Oxford's language monitoring programme, the use of binge-watch increased fourfold in February 2014 and tripled in June 2014.[3]

While the term in this context has yet to make its way into the Oxford English Dictionary, binge-viewing has become such an essential part of the Netflix branding that the company now harnesses the concept for both its production and marketing.[4]

The word binge obviously has a much longer history. Originally referring to the soaking of a vessel, in the nineteenth century it became a slang term meaning to soak oneself in alcohol—that is, through excessive consumption.[5] The word gradually became attached to forms of excessive indulgence, firstly eating, and then other activities.[6] Drawing on the history of the term and its cultural connotations, binge is associated with excess and overindulgence, such that it carries with it suggestions of suspect and even potentially self-harming behavior. Although the lightness of tone with which binge-viewing is often used in popular and scholarly work suggests these negative connotations may well be undergoing a cultural transformation, we nonetheless need to think through the implications of the term's use in our current context. As Debra Ramsay notes, we do not refer to "binge-listening" or "binge-reading," suggesting that the term's adoption for television viewing "implies a vague distaste for the medium itself."[7] Particularly because most scholarship on television bingeing has not interrogated the implications of adopting the word binge from common usage, it is timely that we question the term's role in reflecting and informing the way we currently engage with media. If binge-viewing suggests *excessive* consumption, the question remains, in excess of what?

Rethinking Television Spectatorship

Netflix's association with and promotion of binge-viewing must be placed within the broader history of television, and the shifting dynamics of its relationship with the audience. Although the advent of VCRs, DVD box sets, and now streaming services, offer viewers the ability to concentrate on one program for an extended amount of time, these technologies find their place within the dominant discourse of what John Ellis famously termed the distracted TV glance.[8] This shifting relationship between television and its viewers throughout the medium's history has consistently been framed by reference to older media forms such as radio, cinema, and even the novel. Jay David Bolter and Richard Grusin argue that new media validate themselves through comparison to older established media, situating themselves as a refashioned and improved model in a process that the pair has termed "remediation."[9] Television history, from its early years through to Netflix, is indicative of the way we make sense of the new by reference to the old, but also the conceptual limitations of this approach as we seek to explain the binge experience of Netflix.

Television was initially called "visual radio" in its purely conceptual days, thereby aligning it with the dominant domestic medium of the time.[10] Indeed, radio broadcasting companies were the main investors in television in the developmental years of the 1920s and 1930s. In the post-war years, television was still financed by radio advertising revenue, but once it became financially self-sufficient in the late 1940s and early 1950s, television was explicitly marketed to emphasize its value as a replacement for radio, cinema, and the theater.[11] Lynn Spigel notes how advertisements in the 1950s characterized television in terms of "'family theater,' the 'video theater,' the 'chairside theater,' the 'living room theater,' and so forth," evoking both live theater and the movie theater.[12] Television was marketed as being better than theater, because television could give you a perfect view of the action, and provide the intimacy of the close-up.[13] You did not need to buy the best seats in the theater; you already had them in your home. Like "visual radio," this conception of television stressed the benefits of a domestic medium. The theatrical references, however, suggest that the cultural value of television could only be asserted through comparisons with non-domestic media.

In these early days when television viewing was still seen as a special occasion, John Hartley and Tom O'Regan argue that television spectatorship had a wider social aspect, and what they term a "quasi-cinematic" quality.[14] The novelty of the first person in the street to buy a prohibitively expensive television set meant that neighbors would be invited to watch television as a local community.[15] Over time, as set ownership became more common, this communal viewing gave way to private, family viewing. It was television's status as a type of familiar domestic appliance that led John Ellis to characterize broadcast television as "intimate and everyday, a part of home life rather than any kind of special event."[16] Television went from a deluxe item requiring your full attention, to just another appliance. It was partly due to this domestic context that Ellis suggested television was characterized by a distracted glance.[17] He writes:

> The regime of TV viewing is thus very different from the cinema: TV does not encourage the same degree of spectator concentration. There is no surrounding darkness, no anonymity of the fellow viewers, no large image, no lack of movement amongst the spectators, no rapt attention. TV is not usually the only thing going on, sometimes it is not even the principal thing. TV is treated casually rather than concentratedly. It is something of a last resort ("What's on TV tonight, then?") rather

than a special event. It has a lower degree of sustained concentration from its viewers, but a more extended period of watching and more frequent use than cinema.[18]

In its transition from "visual radio," through to "video theater," and finally to domestic appliance, the remediation of television indicates shifting terms of comparison as well as associated shifts in the audience relationship to the television.

From one perspective, the model of the distracted television glance within the domestic setting might be seen as even more distracted in the current era. Thus *House of Cards* actor Kevin Spacey, presenting the James Mac-Taggart Memorial Lecture at the Guardian Edinburgh International Television Festival in 2013, characterized the shift in television viewing this way:

> Now when I think about what the MacTaggart lecture was like 40 years ago.... I imagine that the audience ... probably went home ... and shared that time-honored tradition when the entire family would gather around the television set.... Today, when I think about how all of you might go home.... It's more likely you've already recorded *It's a Wonderful Life* (1946) on your DVR, as you gamely try to gather the family around the giant movie screen that you've installed in what used to be the garage; then you can try to find out where your children are on Facebook; you might ask your partner to stop Instagramming photos ... of the meal that they've just ordered ... while Grandma desperately pins even more pictures of cats on her Pinterest page, as your son quietly and surreptitiously clears his entire browser history, and your daughter Tweets how boring *It's a Wonderful Life* is ... you too will feel that warm family glow of precious time when we all come together to basically ignore each other.[19]

If we take Spacey's perspective, we might see ourselves as hyper-distracted, multi-tasking television viewers. Of course, Ellis's notion of the distracted television viewer was always inherently one of the multi-tasker who might be engaged in other domestic activities in between—or while in the process of—watching television. The difference, then, is not so much the multitasking itself, but rather the simultaneous multitasking across various media platforms. As Amanda D. Lotz argues, "We may continue to watch television, but the new technologies available to us require new rituals of use."[20]

Spacey's imaginary contemporary family audience calls to mind what new media theorist Dan Harries has referred to as the blurring of "old media *viewing* and new media *using* as 'viewsing.'"[21] It also reflects what Stephen E. Dinehart has called the viewer/user/player, or VUP, for although Spacey frames his multitasking family within a discourse of distraction, it is also possible—and indeed increasingly likely—that the other media we might be using while watching television could be related to the series we are in the process of watching.[22] This extends the nature of our engagement with the series even as it diverts our attention from being focused solely on the television text. The television glance may still be in operation here, but it is a far more complex in the new media context.

Against the concept of the hyper-distracted contemporary television viewer, we must—at one at the same time—allow for its near opposite, the intensely engaged binge-viewer. The sketch-comedy series *Portlandia* (2011–) illustrates this intensity take to ridiculous extremes in "One Moore Episode" (2012), in which a couple become so entranced with the rebooted *Battlestar Galactica* (2004–2009) that they lose their jobs and suffer all manner of physical discomforts—including crusty eyes and bladder infections—all in the name of continuing their binge-viewing. This particular form of binge engagement with television is, then, more focused, more intense, more all-consuming than its cinematic counterpart. Even if we only take into account the practical element of duration of sustained engagement—that is, the total running time that we devote to viewing in one sitting—the idea of television being characterized by a distracted domestic glance is reconfigured in this context. The domestic setting allows us to devote so very many continuous hours to the one program. We are particularly able to binge on our favorite programs at home because we can be comfy in our jammies in bed, or allow ourselves to become crusty and disgusting—as in *Portlandia*—without anyone seeing. We can also both time-shift and space-shift our viewing to other locations, and in the case of Netflix, very easily pick up where we left off.

Many commentators both in the popular press and scholarly writing have noted that binge-viewing changes how we *experience* television, and have reframed this form of television spectatorship in reference to other media. Michael Z. Newman and Elena Levine argue that "the intensity of uninterrupted viewing makes the experience of a TV show more like that [of] a book or film."[23] This common comparison, and the ongoing process of television remediation more broadly, is an attempt to grapple with shifts in spectatorship, but we should be wary of completely dismissing the discourse of distraction. While our *Portlandia* couple becomes incredibly resentful that anyone would want to contact them while viewing *Battlestar*, they still manage to order take-out, and text a friend to get out of a social event. The intense absorption of the *Portlandia* couple in the television text, and their unwillingness to be distracted from it, does not preclude other domestic, social, and media distractions to their viewing, even as they prioritize the television to a ridiculous extent. What we have today, then, is both the hyper-distracted glancing viewer and the focused binge-viewer enabled by co-existing technologies at the same cultural moment, even potentially in the same viewer.

Before DVD box sets and streaming services were even available, fan engagement with television texts suggest that the medium has always been capable of attracting our focused attention (and not just our casual glance), depending on our level of interest in a particular program. Matt Hills argues that the advent of DVD box sets "enable[d] fans to set their own pace of consumption, 'bingeing' on a series by watching multiple episodes at any one time if the narrative pull is sufficient."[24] Yet as Lotz argues, broadcast television, cable, or DVD require "viewers ... to negotiate contradictory fan motivations—viewing as soon as available versus" viewing all at once.[25] This is where Netflix steps in; its marketing strategy very much centered on resolving that conflict. In the contemporary era, bingeing in particular need not be characterized as a fandom practice per se any more but simply as a contemporary viewing norm. In *Portlandia*, binge-watching becomes associated with an obsessive form of fandom, as the couple seeks out an unsuspecting man in their city who happens to share the name of *Battlestar*'s executive producer and writer Ronald D. Moore, in order to press him into writing more episodes. The *Portlandia* couple's obsession with *Battlestar* only dissipates when a new program is found to be binge-worthy—*Doctor Who* (1963–1989, 1996, 2005–).[26] If the majority of viewers now engage in binge-viewing, it becomes an example of the argument mounted by Hills, Henry Jenkins, and Joshua Green that marginal fan activities tend to become mainstream over time, and indeed are increasingly encouraged as a promotional strategy.[27] The ongoing transition towards normalization is important to the way in which we conceptualize binge-viewing—and indeed fandom—because associations with excess become more difficult to sustain. Who, then, is to say what constitutes excess in terms of television viewing?

Epic-Viewing

Netflix markets itself as "the future of television," situating itself as a new form of television that goes beyond the previously dominant temporal norms, such that excess is positioned positively.[28] In her work on the epic in cinema and television, Vivian Sobchack suggests that "an excess of temporality finds it form in, or 'equals,' *extended duration*," of going beyond the normal running time.[29] While Sobchack specifically focuses on the epic as a genre, she also claims that "what *signifies* temporal excess is not universal but culturally and historically determined."[30] Netflix's strategy of designing and releasing texts such as *House of Cards* based upon the assumption and encouragement of consecutive viewing has become one of the new signifiers of temporal excess in Sobchack's terms.

From May through June 2015, Samsung Australia put out a joint promotion for its UHD TVs with a six-month complementary Netflix subscription. The broadcast television and YouTube advertisement depicted a couple who, like Frank Underwood of *House of Cards*, loathe "the necessity of sleep," as they press on for "just one more episode."[31] On HBO, if you want one more episode of *Game of Thrones* (2011–), you have to wait another week. If you want one more episode of *House of Cards*, you can just keep watching. Netflix's marketing and delivery model is thereby based upon an assumed and encouraged mode of consumption that exceeds the temporality of its competitors, and consolidates a new temporal norm for television. With Amazon Prime making *Transparent* (2014–) available all at once "à la Netflix," and NBC "taking a cue from Netflix" by releasing a full season of *Aquarius* (2015–2016), both streaming and broadcast television providers are starting to copy this model, but tellingly in popular discourse it is always with reference to the game changer, Netflix.[32]

Because Sobchack's primary frame of reference is the epic film, she suggests that the genre's transfer to television in the form of the miniseries in the 1970s and 1980s constituted a "formal debasement of the genre."[33] Free broadcast television, she argues, created a medium with "lowered expectations."[34] For Sobchack, the broadcast miniseries formally alters the "temporal field" of the Hollywood historical epic, into an "episodic and fragmented exhibition."[35] It is a decline in spatial terms of the screen size; in temporal terms of the length and cohesion of the text; and in terms of spectatorship. The history of television's response to the cinematic epic in fact reaches back to the 1950s and 1960s, as children's programs and one-off specials attempted to harness the popularity of the epic, albeit within the considerable limitations of television.[36] From this historical, medium-specific vantage point, the miniseries of the 1970s and 1980s was, rather, an expansion of scale.[37] Regardless, Sobchack suggests that television is unable to match the experience of the cinematic epic, arguing that the epic film demands an act of spectator endurance that heightens one's awareness of bodily temporality.[38] She writes:

> On the one hand, experiencing this extraordinary cinematic duration, the spectator as a body-subject is made more presently aware than usual of his or her bodily presence—indeed, is "condemned" to the present and physically "tested" by the length of the film's duration. On the other hand, however, *enduring* the film in the present imprints the body with a brute sense of the possibility of transcending the present, of the literal and material capacity of [a] human being *to continue* and *last through* events.[39]

As Sobchack is keen to point out, our human awareness of our own temporality is always "culturally encoded," and therefore subject to variation within and across cultures and time periods.[40] In our current era, heightened temporality is sent up in both the *Portlandia* skit and the Samsung Netflix promotion, a feat of television endurance in which just one more episode is never enough. For the 70 percent of U.S. viewers watching television successively, surely by comparison it is the cinema—regardless of its visual scale—

that begins to feel very small in temporal terms, its narrative and character development squeezed into only two to three hours running time.[41] The "elevation and transcendence of individual temporality" in Sobchack's terms thereby finds new and even extended expression in successive television viewing.[42] While Sobchack focuses on the epic as a genre with specific thematic and iconographic features, *epic-viewing* in the contemporary era is a mode of consumption that transcends genre, "an excess of temporality" that is manifested across a vast range of program types as we experience them in one, epic expanse.

While heightened temporality may well be a feature of all television bingeing, it is particularly pertinent for the Netflix production model that assumes and encourages a continuous streaming session. We can already indulge in an act of viewer endurance across an extended period time with any program of any genre, using DVD box sets, cable marathons or recorded content, streaming "box sets," or Netflix's considerable licensed back-catalog.[43] Netflix's delivery makes the process easier to both commence and continue, either in the one sitting, or in consecutive sittings, with the ease of picking up where we left off across multiple media platforms. In this regard, streaming services such as Netflix offer a far more streamlined textual and temporal experience than earlier media forms. Netflix has set up an alignment between this extensive viewing experience and the immediacy of its brand new, high-end content. While its competitors weigh the merits of following their lead, currently Netflix is able to offer a number of new programs in full seasons, and then use its extensive library to keep viewers as a secondary branding point. Thus in the lead-up to the launch of Netflix in Australia in March 2015, all advertising material featured *House of Cards*. Upon going to the Australian Netflix website, however, the home page instead emphasized the depth of the library holdings (even though these are notably smaller than Netflix in the United States).[44] This duality exemplifies the fact that Netflix chief content officer Ted Sarandos calls *House of Cards* its "brand ambassador," designed to draw in new viewers, while the secondary appeal of the licensed holdings is intended to keep them.[45] Netflix primarily emphasizes its own premium titles because it is around them that they can stress the appeal of both instantaneous and expansive viewing pleasures of a new, high-end product.

Sobchack's phenomenological focus on the experience of spectatorship helps to point to another aspect of binge terminology. The fact that the term binge has gained traction in the popular press and scholarly work speaks not only to hierarchies of taste situated around different media, but also to the sometimes conflicted feeling of television overload (when we start finding it difficult to track and process what we have seen), and the desire to keep watching to find out what happens next (making it difficult to turn off). These conflicting experiences and feelings surrounding binge-viewing partly account for the numerous terms currently in circulation to describe and explain the phenomena, of which binge is the most prevalent.[46]

Lotz prefers to discuss "successive" viewing, certainly a far more neutral term.[47] It is perhaps too understated, as it describes the practice but not the experience. In colloquial use, the alternative term marathon viewing has been used, suggestive of a lengthy act of endurance on the part of the audience— a challenge that the viewer meets and overcomes rather than one that overcomes the viewer.[48] A marathon implies effort, exhaustion, and perhaps even pride. Does a television marathon suggest something to be boasted about, rather than a guilty pleasure to be confessed? It might even imply that viewing can become hard work as well as—or even rather than—a pleasure. Sobchack stresses that it is the extended duration of the epic that punishes its audience with a feat of physical endurance, but that this very endurance becomes one the genre's pleasures, and contributes towards its creation of a sense of time and history.[49] Taken *beyond* the confines of the epic *genre*, epic-viewing— particularly of expansive texts designed specifically to be viewed over several consecutive hours—thereby shares with marathon viewing this emphasis on endurance. The term television marathon has also been used to refer to a *scheduling* practice used by cable television from the mid–1980s, involving a sequential block of reruns from a particular series, although it has also been employed to refer to thematic block scheduling of programs or films (such as around a particular genre).[50] Marathon viewing comes with quite a different set of associations that are at one remove from either the potential cringe of the binge or the overly neutral phrase successive viewing, but given its history as a scheduling practice and its crossover with film programming, it requires some careful clarification if used. Furthermore, while it speaks to the viewing practice, it does not necessarily imply a particular type of text optimized for this practice. Therefore, the concept of marathon viewing is but *one aspect* of the broader concept of epic-viewing.

Indeed, the "excess of temporality" that Sobchack discusses is not merely a function of viewing duration, but rather epic expansiveness is also created through the materiality of the text's construction and promotion. Bearing this in mind, another colloquial term in circulation—cannonballing—also fails to encompass the broader means for creating this temporal field.[51] Cannonballing evokes either a sense of *speeding* through a series, or jumping into it. A marathon brings to mind a race, but its defining features are its length and endurance rather than speed. Cannonballing, then, ostensibly describes the same activity but with an emphasis on getting through the one series as quickly as possible in the smallest number of sittings, rather than stressing the temporal expansiveness of the text and our experience of it. Marathon viewing and cannonballing each suggest different perceptions of temporality while engaged in extended viewing, and yet there is something lacking in their relationship with the way that the text itself has been both structured and marketed.

Repurposing Television Conventions

Making its in-house seasons available all at once, Netflix organizes its program structure and delivery to channel and prioritize a particular form of viewing experience based on the assumption that a high percentage of its customers are going to continuously watch multiple episodes. Most notoriously, in 2012 Netflix set up the "post-play" default system to automatically keep playing from episode to the next.[52] The end credit sequence, traditionally a paratext that marks the end of the text to lead into the next, is cut short so that you can keep watching.[53] The credits are there if you want them, but they are deemed unnecessary (to the irritation of some customers who felt the feature spoilt the pace and mood). As a result of the backlash, in January 2014 Netflix enabled viewers to turn the feature off if they desired. Post-play is simply one of a string of methods to keep us watching. Before VHS, DVDs, time-shifting technologies, and full-season streaming, "narrative pull" would be designed to keep viewer interest from one week to another, including the use of formal devices such as a cliffhanger ending to an individual episode or a season, to make viewers come back.[54] In an era when viewing can be what Lotz refers to as "deferred" and/or "successive," these same cliffhangers might also prompt a viewer to simply keep watching in the same session, particularly if post-play is still enabled.[55] In this case, a traditional structuring device has simply been reapplied to fit streaming, while others fall by the wayside. This is not a profound departure from the formal structures currently being used by all manner of serialized productions. Rather, cliffhangers and narrative pull are relied upon to achieve a related yet slightly different outcome: to make Netflix's model of continued viewing in the *same* sitting appear more desirable than continued but *delayed* viewing the following week for new content available elsewhere.

Netflix encourages customers to think of its products as best experienced in their *overall* epic duration. In the early days of broadcast television, sustained viewing meant watching a succession of different segments (including advertisements and news breaks) as part of what Raymond Williams famously termed television "flow."[56] Thus Williams argued that watching television actually meant watching television flow, not watching a particular program.[57] Contemporary binge-viewing, by contrast, tends to be associated with watching one episode after another of the same series. Hills suggests DVD box sets separate the television program as individual text, removed from this broader context of traditional broadcast flow.[58] Box sets are thereby able to isolate the viewer's attention. A post-play enabled season on Netflix is similarly able to maintain viewer focus (although at the end of a season it will nonetheless suggest another program to move on to). A television program designed for ad-free streaming has no need for mini-cliffhangers before an advertisement break, and the way traditional episodic structure fits into broadcast television flow becomes irrelevant. Original Netflix content need not fit into a broadcast schedule (as one marketing tagline proclaims, "watch TV on your schedule"), and does not need to have stable episode or season lengths.[59] The heightened association between Netflix's in-house productions and binge-viewing rests partly upon its ability to optimize the structure of its series to fit its delivery model, minimizing repetitive exposition and maximizing what we might think of as a form of streaming flow. This form of optimized streaming flow encourages us to view (conceptually and literally) the television text as an expansive whole. If, for Sobchack, the broadcast miniseries fragments and thereby *reduces* the epic and its overall "temporal field," then I would suggest that the cultivation of binge-viewing through the formal and technological devices employed by Netflix shifts its temporal field *outwards* once more.[60] Contemporary epic-viewing is predicated on the (encouraged) perception of television texts in their overall, epic entirety.

Catering to this form of viewing is increasingly becoming a financial imperative, resulting in changes to television conventions. Epic-enabled series not only appeal to the majority of U.S. viewers who now watch television in successive episodes, but also facilitate the movement of a program from an initial weekly broadcast, to later continuous viewing.[61] As a result, many writers for network and cable television now attempt to structure their series to cater for both types of viewing—weekly and continuous.[62] This is particularly important to secure a secondary market, including licensing through providers such as Netflix. In practical terms, this means being ever more vigilant about continuity, given that errors stand out more in successive viewing. As Jason Mittell argues, "bingeing on DVD can highlight narrative redundancies designed for weekly viewers," such as recaps at the beginning of episodes, and explanatory exposition within episodes.[63] They may, therefore, become either annoying repetitions, or welcome signposts if we have lost concentration along the way.[64] It is financially advantageous for contemporary producers and writers to negotiate these competing demands, but Netflix is able to avoid this dilemma through its immediate delivery of its in-house products.

The Netflix model of the epic-enabled program does not involve a radical departure from traditional television structuring devices, but rather their reapplication to continued streaming. This is evident in *Arrested Development*'s shift from Fox to Netflix. Seasons one through three of *Arrested Development* on Fox featured long-running gags that ran across episodes and seasons and therefore required an attentive viewer.[65] Nonetheless, by comparison Mareike Jenner suggests the series embarked on a more free-form narrative structure when it moved to Netflix for season four, "often creating mini-cliffhangers in the middle of a scene that are not resolved until several episodes later."[66] Although Mittell notes this long-form joke structure in the earlier seasons, the Netflix season four represents a heightening of the device in the reformulation of its entire episodic logic. Through season four, we gradually learn the reasons for the Bluth family's absence from matriarch Lucille Bluth's (Jessica Walter) trial, in a series of overlapping narratives and timelines rather than in a linear progression. As this structure becomes clear, the viewer soon realizes that each episode viewed in isolation does not fully reveal its significance and the way it fits into the other narratives, creating an ongoing puzzle we are invited to put together across the season as a whole. Jenner suggests that this structure of interspersed mini-cliffhangers was intended precisely "to encourage 'binge-viewing'" across multiple episodes.[67]

While Sobchack suggests that the epic requires a feat of spectator endurance—through which our experience of bodily temporality becomes heightened—season four of *Arrested Development* is structured to reward this endurance not only through the narrative puzzle as a pleasure in its own right, but also in its contribution to the show's genre-specific pleasures—that is, the way that the narrative special effect underpins the comedic "coincidences" across the series.[68] The Netflix season of *Arrested Development* becomes the most recent form of what Mittell has termed contemporary "narrative complexity" in television that reflects "a shifting balance" between "episodic and serial forms."[69] Season four adopts the puzzle narrative structure, but also employs what Mittell terms "the narrative special effect. These moments push the operational aesthetic to the foreground, calling attention to the constructed nature of the narration and asking us to marvel at how the writers pulled it off."[70] Given that Mittell cites the intersection of plot strands in earlier seasons of *Arrested Development* by way of example, we should see Netflix season four as a further amplification of the narrative special effect, restructuring the program in its formation of character-specific episodes with overlapping timelines that we can keep track of more easily if we watch them in close succession.[71] The viewer who consumes season four of *Arrested Development* in Netflix's encouraged and preferred continuous mode is thereby rewarded for their intense investment of time by being better placed to piece together its complex overlapping narrative and temporal web, even as they may struggle to both physically endure the experience and mentally maintain concentration (an issue about which series creator Mitch Hurwitz has since expressed concerns).[72]

A desire to reformulate the narrative structure of episodes also partly explains why Kevin Spacey, along with his fellow executive producers, ended up taking *House of Cards* to Netflix. David Fincher, Beau Willimon, and Spacey had tried taking the concept to the other major networks, but every one of them demanded a pilot before going ahead. Spacey and his colleagues wanted to avoid the level of exposition and arbitrary cliffhangers that the pilot format requires, preferring the extended time frame of the series as a whole to let the characters and plot lines play out more gradually.[73] Netflix purchased the series without a pilot, basing their decision instead on their audience data. This is a narrowcast niche audience approach rather than a broadcast mass audience approach of the past, albeit with Internet-enabled broad distribution potential.[74] While Netflix approved the slower narrative exposition for *House of Cards*, the series still leaves the ramification of key events (such as a murder at the end of season one) to the *following* season to play out, an old-school cliffhanger method to bring viewers back. The Netflix approach thereby means texts can be designed to be tighter in their mini-mizing of exposition, and yet more freeform in their pacing, while still using (or reworking, in the case of *Arrested Development*) traditional narrative devices such as the cliffhanger.[75] We should ultimately be wary of considering these narrative strategies as particularly new when viewed in the context of the narrative complexity of cable programming, exemplified in older programs such as *The Sopranos* (1999–2007) or *The Wire* (2002–2008). Encouraging viewers to return for future episodes or seasons is standard. Encouraging viewers to watch a whole, uninterrupted season is epic. Put simply, why wait?

Aesthetics of Expense

The reconfiguration of the television text to fit the Netflix model, and both cultivate and reward the endurance of its viewers, has also become part of an ongoing discourse around hierarchies of quality and taste, and the var-ious ways in which these values may be deliberately promoted as a branding exercise. Timothy Todreas has argued that in the era of digital television we have seen a "great value shift from conduit to content."[76] With the renewed attention paid to content, comes the focus of selling individual titles along the lines of what has variously been described as high-end television, and more often as quality television. Although this latter term has a rather com-plicated history that precedes the current streaming era, for our purposes here it is worth noting that there has been an intersection between the dis-courses of quality TV and perceived binge-worthiness.[77] Among its many debated characteristics, quality television has been associated with high-budgets, critical acclaim, and also with a programming and marketing strat-egy to pitch television to a desirable, high-income audience, particularly viewers who might not otherwise watch the medium.[78] Mario Klarer writes that "unlike one or two decades ago, today almost every single network or premium channel" wants a "flagship" high-end serial drama "in their port-folios," not only to draw a particular demographic but also to raise the status of their brand.[79] The Netflix in-house productions are premised on this brand-building marketing strategy. Thus Spacey, perhaps somewhat immodestly, sees *House of Cards* as belonging to a current "Golden Age of Quality Tele-vision."[80] As Newman and Levine argue, it is only certain types of programs that are identified as quality in each supposed "Golden Age."[81] Bearing this qualification in mind, we might see quality TV in terms of a type of market-driven (high-end) production and promotion strategy that may be present across different eras, even if it enters periods of heightened use. Because texts must be deemed worthy enough to warrant sustained attention, continuous viewing has become particularly linked with the notion of "quality TV."[82]

Netflix has used the immediate delivery of its flagship products (such as *House of Cards* and *Orange Is the New Black* [2013–]) as its point of branding difference in a competitive market. Netflix's original content is in the minority when compared to its far more numerous licensed items, which viewers could also be watching continuously on demand. Despite this, advertising focuses primarily upon Netflix original products as part of the trend towards pre-mium brand creation. Given that other providers also have high-end original products, Netflix tries to position itself as intrinsically superior because its premium products can be enjoyed all at once, immediately, in one grand, expansive, epic-viewing experience. Even as its competitors tentatively start to copy this model, the public discourse surrounding such releases is always tied back to Netflix, and as such Netflix currently maintains its public profile as the home of this new model of delivery and spectatorship.

Sobchack notes that while extended duration is the primary means through which the epic employs excess temporality, there are nonetheless many ways in which temporal excess itself can be signified.[83] The title sequence to *House of Cards* features 37 time-lapse photography sequences of Washington, D.C., by photographer Drew Grace, chosen from 120 shots taken in High Dynamic Range (HDR, which can register the large variations in luminosity between day and night shots) over a six-month time span.[84] Time-lapse photography captures an expanse of time and compresses it, its mastery over time being the site of its formal and aesthetic pleasures. John Ellis notes that while network programs are frequently reducing titles to a mere title card, more expensive cable productions use an extended title sequence as an overt marker of quality.[85] For a Netflix full-season epic experience, a long, repeated title sequence each episode would seem largely redundant, as the viewer does not need to be reminded that their series is starting, and may simply skip the opening sequence.[86] Titles nonetheless continue to contribute to quality branding, and thereby prime our anticipation of the text even into a new era in which their role might otherwise be undermined. The *House of Cards* time-lapse images embody the investment of time and money that produced them, and thereby assert the worthiness of the program as a whole to receive our own investment of time and money.

The title sequence to *House of Cards* also raises an expectation of on-location shooting, and therefore speaks both to the (high) budget of the series and the perceived (high) quality associated with it. D.C. is shown from the Capitol to barrels of waste dumped on the river's edge; a seat of political power along with its discarded filth, presented with the same beautifully lit, crisp HDR photographic aesthetic that manages to make waste look good. The images mirror the political drama in which the ability of the Underwoods to manipulate appearances proves more powerful than the less palatable realities of their lives. It is a landscape that speaks to the literal eradication of people in the titles and the narrative—as people captured in the time-lapse shots were painted out of the title sequence to create the right tone.[87] For season two, this sterility is amplified in the subtle shift to winter images with bare trees, a harshness of environment suited to the Underwoods' increasingly ruthless machinations throughout the season.[88] By drawing on the genre associations and thematic concerns of a particular program, a full title sequence such as *House of Cards* (as opposed to its literally poorer cousin, the brief title card) can become part of our viewing ritual that conspicuously asserts its quality in its images and sounds as a form of pleasure.

As part of this imagery of expense, we should also add star persona, as Oscar-winning actors such as Kevin Spacey in *House of Cards* or Maggie Smith in *Downton Abbey* (2010–2015) are overtly paraded markers of quality.[89] Attracting award-winning actors from the cinema into high-cost television productions helps foster the perception of a "cinematic TV" hybrid, and add brand equity.[90] Indeed, Brett Mills notes that the term "quality TV" is often found alongside cinematic TV, tending to equate simply with a program that looks expensive, in that imagery "is foregrounded as an element of a programme that audiences are invited to take pleasure in."[91] The concept of cinematic TV is one that has gained traction among scholars and the popular press, and yet is also coming under increased criticism.[92] As with the concept of quality TV, Mills points out that "it's clear that the term 'cinematic' is one associated with hierarchical ideas of quality, and is perceived to be a compliment when appropriated for television."[93]

With this focus on the overt markers of high-end production as signs of both aesthetic and cultural value, David Carr of the *New York Times* proclaimed "the vast wasteland of television has been replaced by an *excess* of excellence" that viewers can have difficulty keeping up with.[94] Carr argues that this overwhelming supply of (what he calls) quality television has brought with it "intellectual credibility" such that you don't feel culturally guilty for watching television, rather you feel embarrassed if you have not kept up with the latest paragon of excellence.[95] While this suggests a shift in the media hierarchies of taste, our continued use of the terms cinematic and binge in relation to these texts tells a different story, a lingering from a previous era, in which the cinema is positioned above television, and sustained viewing of a television text is given the same term used for guilty, unhealthy overindulgence in junk food or alcohol.

Netflix has ostensibly sidestepped this hierarchical approach as well as the negative connotations of "binge" by marketing its form of program delivery as "the future of television."[96] This elevation of streaming television contrasts notably with HBO's former tagline as "not TV" but (by implication) something better.[97] In trying to explain the distinguishing features of the Netflix format and mode of delivery, its creative staff nonetheless repeatedly falls back on comparisons to older media forms. *Arrested Development* creator Mitch Hurwitz says the series' shift to Netflix means "what it had become was a novel."[98] Similarly, *House of Cards* showrunner Willimon compares the program to a novel, which can be read in one hit or dipped into at will.[99] Such comments form a continuation of the remediation of television that has been in play since its inception, but also indicate an attempt to conceptualize the oscillation between grazing and intense viewing in the current era. The use of terms such as novelistic and cinematic nonetheless circulate within discourses of quality and taste that can be seen partly as an attempt to recuperate the negative tinge of the term "binge-viewing," and more broadly the practice of continuous television viewing itself. In the specific case of Netflix's in-house products—texts designed at the outset to be enjoyed as one very long, post-play epic text—the perceived (and asserted) qualities of the text itself and our particular rituals of engagement with it intersect. Both the epic-enabled text, and our sustained engagement with it are positioned as the premium "future of television."

Netflix has made the connections between its extended format, novels, and the cinema particularly overt in the promotion of the loosely historical epic *Marco Polo*. Creator John Fusco suggests "this is basically a ten hour long movie.... It's like literature, it's a very novelistic format."[100] These connections appear to be part of a deliberate marketing strategy, as they have been repeated in other interviews with various production members. Although *House of Cards* functions as Netflix's main flagship program, its foregrounding of an aesthetic of expense is reserved in comparison with the fanfare surrounding *Marco Polo*, which has been interpreted as a branding exercise for Netflix's expansion into the international market.[101] Netflix spent a reported $90 million on the ten-episode first season of *Marco Polo*, and promotional interviews surrounding the series have emphasized the scale and expense of the project. Netflix chief content officer Ted Sarandos, along with the series' producers, stressed the 800-strong crew who speak 26 languages; 160 crewmembers in the art department; 400 in construction; and "the sets ... including 51 sets in Malaysia," which "required 130 tons of plaster and 1.6 tons of silicone."[102] This emphasis on "size and scope" is nothing new to the historical epic.[103] Sobchack argues that "the genre *formally repeats* the surge, splendor, and extravagance, the human labor and capital cost entailed by its narrative's *historical content* in both its *production process* and its *modes of representation*."[104] The cost and scale of the production are emphasized in the promotional material and in the visual construction of the text itself, such that the "*history of production* and the *production of history*" become intertwined.[105] Writing in 1990, Sobchack suggests that the labor invested in the television epic miniseries tends not to be publicly celebrated as a marketing strategy.[106] This is clearly no longer the case. The promotion surrounding *Marco Polo* invites us to be impressed by, and take pleasure in, the expense and expanse of the image, the ten-hour running time it takes to display it as well as the work entailed in producing it. Thus, as Sobchack argues, the "temporal excess" of the epic becomes "encoded as empirically verifiable and material excess—entailing scale, quantification, and consumption in relation to money and human labor."[107] This is not to suggest a return to the dubious notion of cinematic television, but rather to note the way in which the concepts of the cinematic and the epic are deliberately employed in the promotion of *Marco Polo*. Reviews have not been favorable, and references to "B-movie clichés" must particularly sting given the considerable expense showered on the program.[108] For our purposes here, what matters is not the success or reception of *Marco Polo*, nor indeed its high-end treatment of quasi-historical subject matter, which we have seen before (in programs such as *Rome* [2005–2007]). Rather, what is important is the way in which Netflix has framed its promotion of the series, linking the grandness of the scale and cost of the series with the grandness of experiencing the program as one epic whole on Netflix. This is intended to infer that to watch series on other providers is to be impoverished by the temporal restrictions of scheduled viewing, to be held back both by textual fragmentation and the act of waiting.

Indeed, Sobchack notes that the primary object of her study is not so much epic films in themselves but rather the language surrounding public experience of them, including promotional materials and reviews; the cultural and experiential field of the epic's temporality.[109] Sobchack suggests that the extratextual discourse about the production of an epic film is an inherent part of creating and expanding its "temporal field."[110] What is revealed in the way that *Marco Polo* has been promoted is not only a marketing strategy common to the Hollywood historical epic genre that Sobchack discusses, but also a culmination of what Netflix has been trying to harness all along as a brand: the sense of consecutive streaming spectatorship as bigger and better than "ordinary" television viewing, as an event and as an experience, as grand and impressive—as epic-viewing. That is, its appeal to the epic extends beyond the field of genre per se, but rather speaks to a broader branding strategy and the promotion of a particular mode of viewing.

Conclusion

Epic-viewing is predicated on spectator endurance of an extensive text, one that frequently foregrounds its construction (albeit in different ways across different programs) as a site of pleasure posited as worthy of that endurance. Netflix's writing, marketing, and delivery methods particularly encourage viewers to conceptualize its programs as epic texts. As other providers start to copy this model, Amazon Studios's head of comedy Joe Lewis proclaims, "We're actually getting to make up this new form of story-telling…. That's the only way to think about it. I don't think about it as binge-ing. We need to figure out a new word for it."[111] By extending Sobchack's work, I have suggested epic-viewing as the term Lewis is searching for, with its associations of heightened temporality. The epic as *a genre* is, therefore, only one manifestation of epic television viewing. The Netflix model sets up a viewing experience in which *any* television serial or episodic/serial hybrid is conceptualized as one whole, expansive, epic text. Epic-viewing becomes a journey that is physically experienced in extended time, as our bodily fatigue conjoins with, and partly underpins, our viewing pleasure of this new epic paradigm.

NOTES

1. I would like to express my thanks to Dr. Diana Sandars at the University of Melbourne, Australia, for her comments on this essay.

2. Bob Verini, "Marathon Viewing Is Forcing Showrunners to Evolve," *Variety*, June 19, 2014, accessed June 1, 2015, http://variety.com/2014/tv/awards/binge-viewing-is-forcing-showrunners-to-evolve-1201221668/.

3. "New Words Added to OxfordDictionaries.com Today Include Binge-Watch, Cray, and Vape," *OxfordWords Blog*, August 2014, accessed June 1, 2015, http://blog.oxforddictionar ies.com/press-releases/new-words-added-oxforddictionaries-com-august-2014/.

4. http://www.OxfordDictionaries.com includes more regular updates on current English, whereas the Oxford English Dictionary (http://www.oed.com) is a historical dictionary, and as such tends to be more conservative in its introduction of new words. At the time of writing, binge-watch had been added to the former but not the latter.

5. Rachel Herring, Virginia Berridge, and Betsy Thorn, "Binge Drinking: An Exploration of a Confused Term," *Journal of Epidemiology and Community Health* 62.6 (2008): 476.

6. Jim Lemon, "Comment on the Concept of Binge Drinking," *Journal of Addictions Nursing* 18.3 (2007): 147–148.

7. Debra Ramsay, "Confessions of a Binge Watcher," *Critical Studies in Television*, October 4 (2013) http://cstonline.tv/confessions-of-a-binge-watcher.

8. Matt Hills, "From the Box in the Corner to the Box Set on the Shelf: TVIII and the

Cultural/Textual Valorizations of DVD," *New Review of Film and Television Studies* 5.1 (2007): 41–60; John Ellis, *Visible Fictions: Cinema:Television:Video* (London: Routledge, 1982), 128, 161–162.

9. Jay David Bolter and Richard Grusin, *Remediation: Understanding New Media* (Cambridge: MIT Press, 1999).

10. Erik Barnouw, *Tube of Plenty: The Evolution of American Television*, 2nd ed. (New York: Oxford University Press, 1990), 20, 73.

11. *Ibid.*, 114.

12. Lynn Spigel, "Installing the Television Set: Popular Discourses on Television and Domestic Space, 1948–1955," in *Private Screenings: Television and the Female Consumer*, ed. Lynn Spigel and Denise Mann (Minneapolis: University of Minnesota Press, 1992), 13–15.

13. Spigel, *Make Room for TV: Television and the Family Ideal in Postwar America* (Chicago: University of Chicago Press, 1992), 23.

14. John Hartley and Tom O'Regan, "Quoting Not Science but Sideboards," in *Teleology: Studies in Television*, ed. John Hartley (London & New York: Routledge, 1992), 206.

15. *Ibid.*

16. Ellis, *Visible Fictions*, 113.

17. See for example Ellis, *Visible Fictions*, 128, 161–162.

18. *Ibid.*, 128. Similarly, Roland Barthes argues that television is "the opposite experience" of cinema. Quoted in Sandy Flitterman-Lewis, "Psychoanalysis, Film, and Television," in *Channels of Discourse, Reassembled*, ed. Robert C. Allen (London: Routledge, 1992), 217.

19. "Kevin Spacey MacTaggart lecture—video," *The Guardian*, August 23, 2013, accessed June 1, 2015, http://www.theguardian.com/media/video/2013/aug/23/kevin-spacey-mactaggart-lecture-video. A full transcript, which is slightly different to the delivered lecture, is also available. "Kevin Spacey MacTaggart lecture—full text," *The Guardian*, August 23, 2013, accessed June 1, 2015, http://www.theguardian.com/media/interactive/2013/aug/22/kevin-spacey-mactaggart-lecture-full-text. The version quoted here refers to the lecture as it was delivered.

20. Amanda D. Lotz, *The Television Will Be Revolutionized* (New York: New York University Press, 2007), 2–3.

21. *Ibid.*, 17.

22. Stephen E. Dinehart, "Transmedial Play: Cognitive and Cross-Platform Narrative," *The Narrative Design Explorer: A Publication Dedicated to Exploring Interactive Storytelling*, May 14, 2008, accessed June 1, 2015, http://narrativedesign.org/2008/05/transmedial-play-cognitive-and-cross-platform-narrative/. See also the discussion of transmedia and VUPs in Angela Ndalianis, *The Horror Sensorium: Media and the Senses* (Jefferson, NC: McFarland, 2012), 172–173.

23. Michael Z. Newman and Elana Levine, *Legitimating Television: Media Convergence and Cultural Status* (New York: Routledge, 2012), 141.

24. Hills, "From the Box in the Corner to the Box Set on the Shelf," 58.

25. Amanda D. Lotz, "Rethinking Meaning Making: Watching Serial TV on DVD," *Flow* 4.12 (2006), http://flowtv.org/2006/09/rethinking-meaning-making-watching-serial-tv-on-dvd/.

26. Thus Matt Hills argues fandom studies tend to be too singular in their focus on fandom of a specific text. A fan might move from one text to the next, or may be a fan of several texts at once. See Hills, "Patterns of Surprise: The 'Aleatory Object' in Psychoanalytic Ethnography and Cyclical Fandom," *American Behavioral Scientist* 48 (2005): 801–821.

27. Bob Verini, "Marathon Viewing Is Forcing Showrunners to Evolve"; Henry Jenkins, *Fans, Bloggers, and Gamers: Exploring Participatory Culture* (New York: New York University Press, 2006), 134–151; Matt Hills, "Defining Cult TV: Texts, Inter-Texts and Fan Audiences," in *The Television Studies Reader*, ed. Robert C. Allen and Annette Hill (London: Routledge, 2004), 509–523; Joshua Green and Henry Jenkins, "The Moral Economy of Web 2.0," *Media Industries: History, Theory, and Method*, ed. Jennifer Holt and Alisa Perren (Oxford: Wiley-Blackwell, 2009), 216, 222.

28. "Netflix Original Series—The Future of Television Is Here," *YouTube*, posted by Netflix, September 3, 2013, accessed June 1, 2015, https://www.youtube.com/watch?v=_kOvUuMowVs.

29. Vivian Sobchack, "'Surge and Splendor': A Phenomenology of the Hollywood Historical Epic," *Representations* 29 (1990): 37.

30. *Ibid.*, 29–30.

31. "Samsung UHD TV Netflix Promotion 1," *YouTube*, posted by Samsung Australia, May 10, 2015, accessed June 1, 2015, https://www.youtube.com/watch?v=-TwCGh8KsPY (video discontinued).

32. Philiana Ng, "'Transparent' Team Talks Binge Viewing, Defends Digital Platform Pay," *The Hollywood Reporter*, July 12, 2014, accessed June 1, 2015, http://www.hollywood reporter.com/live-feed/transparent-team-talks-binge-viewing-718157; Michael O'Connell, "NBC Releasing Complete 'Aquarius' Season on Premiere Day," *The Hollywood Reporter*, April 29, 2015, accessed June 1, 2015 http://www.hollywoodreporter.com/live-feed/nbc-releas ing-complete-aquarius-season-792478.

33. Sobchack, "'Surge and Splendor,'" 41. For a more extensive history of the miniseries, see John De Vito and Frank Tropea, *Epic Television Miniseries: A Critical History* (Jefferson, NC: McFarland, 2010).

34. Sobchack, "'Surge and Splendor,'" 41.

35. *Ibid.*, 42.

36. Djoymi Baker, "'The Illusion of Magnitude': Adapting the Epic from Film to Television," *Senses of Cinema* 41 (2006), http://sensesofcinema.com/2006/film-history-confer ence-papers/adapting-epic-film-tv/.

37. *Ibid.*

38. Sobchack, "'Surge and Splendor,'" 37.

39. *Ibid.*, 37–8.

40. *Ibid.*, 39. Writing in 1990, Sobchack suggests that "in the electronic era of the television and the VCR, temporality is transformed" in that "one can materially and literally manipulate time," a factor that she discusses in terms of postmodern approaches to history. This relates particularly to her broader argument about the historical epic's ability create the feeling of "being-in-History" (rather than merely being in time), a genre-specific argument that I will not examine here for reasons of scope. *Ibid.*, 42.

41. Verini, "Marathon Viewing Is Forcing Showrunners to Evolve."

42. Sobchack, "'Surge and Splendor,'" 29.

43. For example, Australian streaming service Stan advertises "hundreds of series including complete box sets." "Stan. The Biggest Deal in Entertainment," Stan, 2015, https://www.stan.com.au/.

44. At the time of writing, the Netflix Australia tagline had changed to "Watch TV shows & movies anytime, anywhere." Netflix Australia, accessed June 21, 2015, https://www.netflix.com/au/; Ben Grubb, "How the Australian Netflix Differs from the U.S. Service," *Sydney Morning Herald*, March 24, 2015, accessed June 1, 2015, http://www.smh.com.au/digital-life/hometech/how-the-australian-netflix-differs-from-the-us-service-20150324-1m60g8.

45. Quoted in Michael Idato, "*House of Cards* Season 3 Launches on U.S. Netflix, Will Australians Be Watching?" *Sydney Morning Herald*, February 27, 2015, accessed June 1, 2015, http://www.smh.com.au/entertainment/tv-and-radio/house-of-cards-season-3-launches-on-us-netflix-will-australians-be-watching-20150227-13qc1k.html.

46. A new term has even arisen to explain continued viewing of a program we do not like: purge-watching. Adam Sternbergh, "Make It Stop: When Binge-Watching Turns to Purge-Watching," *Vulture*, April 21, 2015, accessed June 1, 2015, http://www.vulture.com/2015/04/when-binge-watching-turns-to-purge-watching.html.

47. Lotz, "Rethinking Meaning Making."

48. Verini, "Marathon Viewing Is Forcing Showrunners to Evolve."

49. Sobchack, "'Surge and Splendor,'" 37–38.

50. Megan Mullen, *The Rise of Cable Programming in the United States: Revolution or Evolution?* (Austin: University of Texas Press, 2003), 167–168; Barbara Klinger, "24/7: Cable Television, Hollywood, and the Narrative Feature Film," in *The Wiley-Blackwell History of American Film*, ed. Cynthia Lucia, Roy Grundmann, and Art Simon (Oxford: Wiley-Blackwell, 2012), 302.

51. The term cannonballing appears not to have been adopted in scholarly work as yet,

but is used colloquially and in the popular press. See for example, Wendy McClure and Maris Kreizman, *"Arrested Development* Binge-Watch vs. One a Week: Which Viewing Strategy Worked Best?" *Vulture,* September 3, 2013, accessed June 1, 2015, http://www.vulture.com/2013/09/arrested-development-viewing-tortoise-vs-hare.html.

52. Jason Gilbert, "Netflix's 'Post-Play' Feature Will Suck You into More TV Show Marathons," *Huffington Post,* August 16, 2012, accessed June 1, 2015, http://www.huffington post.com/2012/08/16/netflix-unveils-post-play_n_1789111.html.

53. See Jonathan Gray, *Show Sold Separately: Promos, Spoilers, and Other Media Paratexts* (New York: New York University Press, 2009), 23–46.

54. Hills, "From the Box in the Corner to the Box Set on the Shelf," 58.

55. Lotz, "Rethinking Meaning Making."

56. Raymond Williams, *Television: Technology and Cultural Form* (Glasgow: Fontana/Collins, 1974), 89–90.

57. *Ibid.,* 89. See also, Gregory A. Waller, "Flow, Genre, and the Television Text," in *In the Eye of the Beholder: Critical Perspectives in Popular Film and Television,* ed. Gary R. Edgerton, Michael T. Marsden, and Jack Nachbar (Bowling Green, OH: Bowling Green State University Press, 1997), 59–61.

58. Hills, "From the Box in the Corner to the Box Set on the Shelf," 45.

59. "Netflix Quick Guide: What Is Streaming And Why Is It Better?" *YouTube,* posted by Netflix, May 21, 2013, accessed June 1, 2015, https://www.youtube.com/watch?v=lW9BL mjzi_4&list=PLvahqwMqN4M3HhXOhAybp03ysb6uxBXUf&index=1.

60. Sobchack, "'Surge and Splendor,'" 42.

61. Verini, "Marathon Viewing Is Forcing Showrunners to Evolve."

62. *Ibid.*

63. Quoted in Lotz, "Rethinking Meaning Making"; Thus Simon Blackwell, co-executive producer of *Veep* (2012–) argues that it would be preferable to adopt binge-viewing as the default position, because it would negate the need for too much exposition referring back to earlier events—you would just assume that your audience remembered. See Verini, "Marathon Viewing Is Forcing Showrunners to Evolve." Elsewhere Jason Mittell discusses the now-famous false teasers for *Arrested Development* that are never shown in the series itself. See Mittell, "Narrative Complexity in Contemporary American Television," *The Velvet Light Trap* 58 (2006): 34. This gag predates the series move to Netflix but becomes more obvious in successive viewing on DVD or streaming.

64. More broadly, Lotz argues that there can be a "profound difference in meaning available to those who watch a season or even an entire series over the course of a few days or even a month," particularly as viewers can forget plot details if watching a complex series over a period of years. See Lotz, "Rethinking Meaning Making."

65. Mittell, "Narrative Complexity in Contemporary American Television," 34.

66. Mareike Jenner, "A Semi-Original Netflix Series: Thoughts on Narrative Structure in *Arrested Development* Season 4," *Critical Studies in Television,* June 6, 2013, http://www.cstonline.tv/semi-original-netflix-arrested-development.

67. *Ibid.*

68. Sobchack, "'Surge and Splendor,'" 37. Despite this, Mitch Hurwitz suggests the rewards for successive viewing may be genre specific, and perhaps not ultimately suited for comedy if a tired viewer starts missing jokes. This had not occurred to him until in-house test screenings of the completed season. The issue of the genre-specific pleasures of successive viewing is one that warrants future detailed consideration. See Denise Martin, "Mitch Hurwitz Explains His *Arrested Development* Rules: Watch New Episodes in Order, and Not All at Once," *Vulture,* May 22, 2013, accessed June 1, 2015, http://www.vulture.com/2013/05/mitch-hurwitz-dont-binge-watch-arrested-development.html. Michael Z. Newman suggests we be wary of this shift in consumption given the loss of connection with weekly viewers watching the same thing at the same time. Michael Z. Newman, "TV Binge, *Flow* 9.05 (2009), http://www.flowjournal.org/2009/01/tv-binge-michael-z-newman-university-of-wisconsin-milwaukee/.

69. Mittell, "Narrative Complexity in Contemporary American Television," 32.

70. *Ibid.,* 38, 35. See also Mittell, "The Qualities of Complexity: Vast Versus Dense

Seriality in Contemporary Television," in *Television Aesthetics and Style*, ed. Steven Peacock and Jason Jacobs (New York: Bloomsbury Academic, 2013), 45–56.

71. It is worth noting, however, that the limited availability of the actors was the deciding factor in the character-specific structure of the fourth season. Lacey Rose, "'Arrested Development' Stars' Surprising Salaries Revealed," *The Hollywood Reporter*, May 22, 2013, accessed June 1, 2015, http://www.hollywoodreporter.com/live-feed/arrested-development-stars-surprising-salaries-526530.

72. Martin, "Mitch Hurwitz Explains His *Arrested Development* Rules."

73. "Kevin Spacey MacTaggart lecture—video."

74. Lotz writes, "The U.S. television audience now can rarely be categorized as a mass audience; instead, it is more accurately understood as a collection of niche audiences" and has changed to "a narrowcast medium—one targeted to distinct and isolated subsections of the audience." See Lotz, *The Television Will be Revolutionized*, 5. It is on this basis that Netflix uses its considerable customer data to develop niche-directed series such as *House of Cards*.

75. While *House of Cards* is predominantly structured to minimize exposition, a noticeable exception occurs when dialog late in the penultimate episode ("Chapter 38" [2015]) of season three is repeated at the beginning of the final episode ("Chapter 39" [2015]), creating a jarring temporal loop when viewed continuously. The predominant Netflix textual model becomes more noticeable in these small moments of irregularity.

76. Timothy M. Todreas, *Value Creation and Branding in Television's Digital Age* (Westport, CT: Quorum Books, 1999), 7. For reasons of scope this article will not delve into the periodization of television history, but a number of scholars chart the transition from TVI (the broadcast era), TVII (sometimes referred to as the cable era), and the current TVIII (the post-network era) in which Todreas locates this shift to content. See Lotz, *The Television Will Be Revolutionized*, and Ellis, *Seeing Things: Television in the Age of Uncertainty* (London: I.B. Tauris, 2000).

77. Robin Nelson uses the term high-end television to denote "big budgets and the high production values associated with them." See Nelson, *State of Play: Contemporary "High-End" TV Drama* (Manchester: Manchester University Press, 2007), 2; Shawn Shimpach, *Television in Transition* (Oxford: Wiley-Blackwell, 2010), 134–135. For reasons of scope, the full history of "quality TV" as a term will not be extensively outlined here, but rather will remain focused on its relationship with Netflix and binge-viewing in particular.

78. See for example Al Auster, "HBO's Approach to Generic Transformation," in *Thinking Outside the Box: A Contemporary Television Genre Reader*, ed. Gary Edgerton and Brian Rose (Lexington: University Press of Kentucky, 2005), 226–227; Jane Feuer, "The MTM Style," in *MTM: Quality Television*, ed. Jane Feuer, Paul Kerr, and Tise Vahimagi (London: British Film Institute, 1984), 34, 56.

79. Mario Klarer, "Putting Television 'Aside': Novel Narration in *House of Cards*," *New Review of Film and Television Studies* 12.2 (2014): 204.

80. John Plunkett and Jason Deans, "Kevin Spacey: Television has Entered a New Golden Age," *The Guardian*, August 23, 2013, accessed June 1, 2015, http://www.theguardian.com/media/2013/aug/22/kevin-spacey-tv-golden-age.

81. Newman and Levine, *Legitimating Television*, 36.

82. *Ibid.*, 141.

83. Sobchack, "'Surge and Splendor,'" 29–30, 37.

84. Alexandros Maragos, "Andrew Geraci Interview. Netflix—*House of Cards*: The Making of the Opening Sequence," *Momentum: Visuals, Aesthetics, Sounds*, February 2013, accessed June 1, 2015, http://www.alexandrosmaragos.com/2013/02/andrew-geraci-interview.html.

85. Ellis, "Whatever Happened to the Title Sequence?" *Critical Studies in Television*, April 1, 2011, http://www.cstonline.tv/letter-from-america-4. Compare also Ellis's earlier work on the broadcast era where he calls the television title sequence "a commercial for the programme itself." See Ellis, *Visible Fictions*, 119–120.

86. Indeed, for some programs, such as Netflix original *Grace and Frankie* (2015–), the post-play function automatically skips the title sequence, except upon returning after a break on another day.

87. Ivan Radford, "10 Things We Learned from the *House of Cards* Director's Commentary," *Vodzilla: The UK's Only Video On-Demand Magazine*, January 5, 2014, accessed June 1, 2015, http://vodzilla.co/blog/features/10-things-we-learned-from-the-house-of-cards-directors-commentary/.

88. For a side-by-side comparison of shots, see David Friedman, "*House of Cards* Season 2 Opening Credits Comparison in Animated GIFs," *Ironic Sans*, February 21, 2014, accessed June 1, 2015, http://www.ironicsans.com/2014/02/house_of_cards_season_2_open in.html.

89. In the United Kingdom, the government has made the link between cinematic television and high-end costs overt, planning tax breaks to what it calls "cinematic television dramas" such as *Downton Abbey*, defined as "those that cost at least £1m an hour to film." Brett Mills, "What Does It Mean to Call Television 'Cinematic'?" in *Television Aesthetics and Style*, ed. Steven Peacock and Jason Jacobs (New York: Bloomsbury Academic, 2013), 64.

90. We might similarly include Netflix's promotion of high profile directors and actors, such as *Sense8* (2015–) creators Lily and Lana Wachowski, or Brad Pitt's involvement in *War Machine* (2017).

91. Mills, "What Does It Mean to Call Television 'Cinematic'?" 64. See also Robin Nelson, who equates the "cinematic look" of quality TV with increased budgets and technological equipment to achieve a comparable visual aesthetic. Nelson, "Quality TV Drama: Estimations and Influences through Time and Space," in *Quality TV: Contemporary American Television and Beyond*, ed. Janet McCabe and Kim Akass (London: I.B. Tauris, 2007), 43, 48.

92. Deborah L. Jaramillo argues, "'Cinematic' should be a contentious word in the field of television studies." See Jaramillo, "Rescuing Television from 'The Cinematic': The Perils of Dismissing Television Style," in *Television Aesthetics and Style*, ed. Steven Peacock and Jason Jacobs (New York: Bloomsbury Academic, 2013), 67.

93. Mills, "What Does It Mean to Call Television 'Cinematic'?" 63.

94. David Carr, "Barely Keeping Up in TV's New Golden Age," *New York Times*, March 9, 2014, accessed June 1, 2015, http://www.nytimes.com/2014/03/10/business/media/fenced-in-by-televisions-excess-of-excellence.html?_r=0.

95. John Ellis similarly argues that on demand television can leave us with "choice fatigue" and "time famine," an anxious feeling of not being able to fit everything in. See Ellis, *Seeing Things*, 170–171.

96. The press has conveniently picked up on this branding. See for example Ken Auletta, "Outside the Box: Netflix and the Future of Television," *New Yorker*, February 3, 2014, accessed June 1, 2015, http://www.newyorker.com/magazine/2014/02/03/outside-the-box-2.

97. Jaramillo, "Rescuing Television from 'The Cinematic,'" 68.

98. Ashley Lee, "NYTVF: *Arrested Development*'s Mitch Hurwitz Wants a Bluth Movie and Season 5 at Netflix," *The Hollywood Reporter*, October 22, 2013, accessed June 1, 2015, http://www.hollywoodreporter.com/news/nytvf-arrested-developments-mitch-hurwitz-650002. Such comparisons are by no means new. Writing in 1999, Vincent Canby at the *New York Times* suggested that surely when conceived as a whole, with its expansive narrative arc, *The Sopranos* should not be seen as a television series at all, but rather a "megamovie" on the scale of epic novels and films. Vincent Canby, "From the Humble Mini-Series Comes the Magnificent Megamovie," *New York Times*, October 31, 1999, accessed June 1, 2015, http://www.nytimes.com/1999/10/31/arts/from-the-humble-mini-series-comes-the-magnificent-megamovie.html. See also Robert Thompson, "Preface," in *Quality TV: Contemporary American Television and Beyond*, ed. Janet McCabe and Kim Akass (London: I.B. Tauris, 2007), xix–xx. More recently, scholar Mario Klarer has made similar claims of *House of Cards*, comparing the series to a novel or "an epic movie." See Klarer, "Putting Television 'Aside,'" 215.

99. Verini, "Marathon Viewing Is Forcing Showrunners to Evolve."

100. Vinnie Mancuso, "Creators and Cast of 'Marco Polo' Discuss Their World-Spanning, $90 Million Show," *New York Observer*, December 11, 2014, accessed June 1, 2015, http://observer.com/2014/12/creators-and-cast-of-marco-polo-discuss-their-world-spanning-90-million-production/.

101. Emily Steel, "How to Build an Empire, the Netflix Way," *New York Times*, Novem-

ber 29, 2014, accessed June 1, 2015, http://www.nytimes.com/2014/11/30/business/media/how-to-build-an-empire-the-netflix-way-.html?_r=0.

102. Todd Spangler, "'Marco Polo' Premiere: Netflix Launches Bid for 13th Century Empire to Conquer the Globe," *Variety*, December 2, 2014, accessed June 1, 2015, http://variety.com/2014/scene/news/marco-polo-premiere-netflix-launches-bid-for-13th-century-empire-to-conquer-the-globe-1201368034/.

103. *Ibid.*

104. Sobchack, "'Surge and Splendor,'" 29; emphasis in original.

105. *Ibid.*, 31; emphasis in original.

106. *Ibid.*, 41–42.

107. *Ibid.*, 30.

108. Brian Lowry, "TV Review: 'Marco Polo,'" *Variety*, November 25, 2014, accessed June 1, 2015, http://variety.com/2014/digital/reviews/tv-review-marco-polo-1201360750/.

109. Sobchack, "'Surge and Splendor,'" 27.

110. *Ibid.*, 35.

111. Lewis's comments about *Transparent* mirror those discussed in this essay around Netflix titles: "It's novelistic; it's not episodic.... We've never looked at this as anything but a continuous piece of five-hour entertainment." Joe Lewis, quoted in Philiana Ng, "'Transparent' Team Talks Binge Viewing."

Streaming Culture, the Centrifugal Development of the Internet and Our Precarious Digital Future[1]

JOSEPH DONICA

> We reject kings, presidents, and voting. We believe in rough consensus and running code.—David Clark to the Internet Engineering Task Force, 1992
>
> There are moments when the quantitative becomes qualitative.—Thomas Piketty, *Capital in the Twenty-First Century*, 2014

In his history of the Internet, Johnny Ryan describes the "defining pattern of the emerging digital age" as the "absence of the central dot." Webs and networks have replaced that center. "This story," he says, "is about the death of the [center] and the development of commercial and political life in a networked system. It is also the story about the coming power of the networked individual as the new vital unit of effective participation and creativity."[2] The problem that early engineers of what would come to be called the Internet tried to solve was the centripetal nature of existing communications systems. Not since Bell and Edison, both of whom worked on centralizing communications based on telegraphy, had communication been so radically reconceived. But with the developments that allowed the Internet to come into existence there was also an idea that communication should be a public utility and one that should operate on centrifugal principles. The idea of a centrifugal communications network as a public utility arose from the citizenry's growing desire to pull away from the center in the post–World

55

War II era. Centrifugal force is simply the force that draws a rotating body away from the center of rotation. This movement away from the center in the development of the Internet has frustrated both big business developments as well as governments—both of whom are only beginning to find ways of subverting the centrifugal nature of the web.

The controversy over Netflix's talks with Comcast that would have caused the video streaming service to begin paying for access to Comcast customers sparked widespread protest and catalyzed conversation about "net neutrality"—the idea that all content providers should have equal access to users and those users should, in turn, have equal access to content. The primary issue on which net neutrality hinges is how governments should treat data shared on the Internet. Supporters of net neutrality want governments to treat all data equally regardless of the user's identity or her status with her service provider. In 2003, Columbia University law professor Tim Wu coined the term in conjunction with his work on "common carriers." Common carrier is a concept in media law that places responsibility on the carrier of a good to ensure its delivery without regard to the party requesting the delivery. Wu argues that the Internet should be regulated just like any other utility— the underlying disagreement he has with supporters of a tiered Internet. The concept is applied to the Internet this way: companies providing the "plumbing" of the Internet, such as cables and modems, should not have the ability to restrict how people use it.[3] Jeff Sommer of the *New York Times* describes the high stakes involved in the debate. According to Sommer:

> The [Federal Communications Commission] has signaled its intention to grant cable and telephone companies the right to charge content companies like Netflix, Google, Yahoo or Facebook for speeding up transmissions to people's homes. And this is happening as the [FCC] is considering whether to bless the merger of Comcast and Time Warner Cable, which could put a single company in control of the Internet pipes into 40 percent of American homes.[4]

The responses from major players in the debate have been mixed, but Facebook's Mark Zuckerberg stressed the coming changes in access by stating, "It's not sustainable to offer the whole Internet for free."[5] The Netflix neutrality case is important because it gives insight into the debates that are shaping the digital future. Netflix is a huge player in the development of that future. The history of the development of the Internet—the rise of Netflix being central to that history—helps us understand the radical changes that have occurred in these modern debates, fundamentally reshaping a previously decentralized and largely unregulated Internet.

Even before Netflix made the deal with Comcast—a deal that the FCC eventually struck down—the company was criticized for the supposed anonymized information it kept about users. Hackers, unsurprisingly, easily found ways to de-anonymize the profiles and access user information. Thus, even before the calls for true net neutrality, Internet activists called for an openness that would still protect user data. But what do policies that ensure that the Internet will remain open to all users while still protecting that data look like? Two high-profile leaks showed the public that government-controlled sites and government use of the Internet do not ensure the protection of user data. This makes it unlikely that policy will eventually side with the protection of all user data. Chelsea Manning's leak of classified information through WikiLeaks and Edward Snowden's release of National Security Agency (NSA) documents to Glenn Greenwald, Laura Poitras, Ewen MacAskill, and Barton Gellman revealed that the citizen's role in the age of big data is tremendously complex. This complexity arises from the fact that citizens, most times unwittingly, release so much data to state agencies and corporations during the course of conducting basic commercial transactions. One need look no further than the terms of service agreements consumers sign just to use new software and complete day-to-day tasks: downloading music and podcasts, filing taxes, or shopping with an online retailer. Some organizations and policy makers are working on open data initiatives that would in some ways counteract the restricted Internet model proposed by Comcast and Netflix.[6] Nonetheless, data is the most valuable commodity in the digital age, and companies have a major stake in acquiring as much of it as they can. Netflix's position on net neutrality provides key insight into the digital future, primarily because Netflix is a major player in that future. But the role the company will play in keeping the Internet the open, democratic—sometimes chaotically so—space it has been since its creation is uncertain. This essay addresses that uncertainty.

Centrifugal force is helpful in describing the development of the Internet as we know it because it defines the decentralized experience of the Internet. Centrifugal force, a concept used primarily in physics and classical mechanics, refers to the tendency of an object moving in a circular motion to be pushed out from the center. The only force on the object is its own gravity, so the greater the size of the object, the further the object will be pushed from the center. Think of a child versus an adult on a merry-go-round. A child spinning on a merry-go-round will be pushed away from the center of it with some force, and an adult on a merry-go-round will be pushed away from the center with even greater force since the adult's weight produces more force away from the center.[7] Translating the concept of centrifugal force into a digital and cultural concept to describe the trends in computer engineering and software development that created the Internet as we know it, I first examine several ways in which an "open Internet" that remains safe for users can develop from within the processes and the structures of the Internet itself. Next, I review the development of the Internet as a centrifugal communication outlet meant to destabilize the control of data in any one center and how this influences a reading of the debate over the deal between Comcast and Netflix that was eventually struck down. Finally, by examining Netflix's corporate culture—a culture that encourages the "hacker ethic" by hosting a hack week at its headquarters, and yet is criticized by the hacking community for undermining the basic protocols on which the Internet was built, I demonstrate the way Netflix is altering the course of the Internet. Although the Occupy Wall Street movement has called for Netflix and Google to take a stand for a more open Internet and the FCC seems to be holding to that standard, the future of the Internet is unclear and largely up to enormously powerful companies like Netflix to determine the outcome.

The Centrifugal Development of the Internet

Surprisingly, there are only a few book-length studies of the development of the Internet. Nevertheless, most try to focus on one aspect of the Internet's development and not the whole history. The reason for the lack of comprehensive histories is most likely the fact that to write a history of the Internet one must take on a subject that has a definite history but one that changes by the minute. The moment one stops writing such a book, it is automatically out of date. As history's institutional goal is to show implications of events, it is still somewhat difficult to determine just how profoundly the Internet is affecting our lives and what extent it will change them in just a few years. However, an aspect of the development of the Internet that appears in most of these histories is the decentralizing ethic of the engineers and developers who built the Internet to be a centrifugal force in modern life. Ryan's history of the Internet explains the importance of this development model as putting "power in the hands of the individual, power to challenge even the state, to compete for markets across the globe, to demand and create new types of media, to subvert a society—or to elect a president."[8] The potential influence of one technology to radically alter society and politics in these ways (unimaginable, perhaps, even to the early developers of the Internet) is an understood part of daily life now. But the Internet's immense influence and power is just now becoming one of the most contested sites of rights, freedom of access, and questions of who should control information—a development that the early Internet's often apolitical engineers would find troubling or at least strange. In the case of Netflix and Comcast, the FCC's ruling to keep the Internet neutral to user access represents one of the first instances that the Internet itself, and not the application of the Internet, is coming under regulatory scrutiny—scrutiny that may change the very structure of how the Internet operates. While the Internet is a centrifugal, user-driven, and open collection of technologies, Ryan says there will be battles and growing pains as the world adjusts to "the new global commons, a political and media system in flux."[9] Those who can compete for "the future of competitive creativity" define this system.[10]

The Internet, as Ryan outlines, was developed in the years just after World War II, a time of tension between the United States and Soviet Union when President Harry S. Truman's strategy advisors were seeking out forms of rapid communication that could transfer unprecedented amounts of information from node to node on existing network connections. The primary concern of his advisors was the communication between sites that housed nukes. In the event of the threat of attack by the Soviet Union, communication between the sites and central command would be crucial, as the failure to inform nuke controllers of a false alarm could lead to nuclear war. His advisors suggested a system that had no central hub. If, during an attack, a central communication hub were destroyed, remote nuke operators would be left without any direction from central command. Their suggested system, first sketched out by Paul Baran of RAND (a U.S. think tank), proposed a system based on the structure of the human brain. His alternative to the existing analog communications system was "a centrifugal distribution of control points: a distributed network that had no vulnerable central point and could rely on redundancy."[11] Baran's goal was to increase communication networks in order to avoid war—not to make it easier to wage.[12]

Ryan's goal in outlining the history of the Internet in this way is to show that the centrifugal principles that the structure on which the Internet was built still guide its protocols and software today. Thus engineers who built the technologies that would eventually come together to become the Internet put power not in any central point but responsibility at the nodal level so that each node became as equal as any other. Baran's concept, and eventually the Internet, became a user-to-user rather than center-to-center communications model. The concept at the time was radical, and when approached by the Air Force to test out Baran's system AT&T dismissed it outright.[13] After AT&T's refusal to test Baran's idea (because of its failure of imagination), the project was shelved until a team could be assembled that had an open vision of what networks could become. As seen in a question Baran posed at the end of a 1962 memo on distributed communication networks, the vision for such a system was about much more than a nuclear-proof network. Baran asked, "Is it now time to start thinking about a new and possibly non-existent public utility, a common user digital data plant designed specifically for the transmission of digital data among a large set of subscribers?"[14]

The history of the precursor to the Internet, ARPANET (the network developed by the Department of Defense's Advanced Research Projects Agency), has been detailed Ryan's work, and by others. Here, though, I want to point to a few of the features in the development of ARPANET that led to its openness and decentralized structure. First, ARPA had a culture in its laboratories of liberality with research funding and little constraint on needing any application to the technologies its researchers developed. Secondly, the early protocols (codes that enable machines on networks to communicate with each other) that came to define the Internet were informal and run by consensus. Steve Crocker developed the first set of protocols just a year out of his undergraduate studies at UCLA. He developed them, he later recalled, in the bathroom so as not to disturb the sleep of other ARPA developers with whom he was sharing an apartment. These protocols would come to define the way machines communicate over the Internet for the next half-century.[15] In a memo sent shortly after Crocker wrote the first protocols, he set the open tone for their future development. The memo stated, "Closely related to keeping the technical design open was keeping the social process around the design open as well. Anyone was welcome to join the party."[16]

Even with the government's interventions and funding in developing the infrastructure of the Internet and funding large portions of the research that went into it, the engineers were able to keep the culture of the Internet one of openness. Their driving virtue was that of universal access. Arguably, this openness and access eventually drove down the prices of machines that connect to it. The government as well as the corporations that would grow out of these early developments relied on collaboration with generally leftist engineers and designers as well as the full-fledged techno-anarchists with activist agendas. Such collaboration brought forth social computing and has become the primary use of the Internet decades later. One of the major motivations for early engineers to socialize computers over a network was so gamers could connect with each other easily. As ARPA made deals to expand its network to nodes with specific interests, such as CSNET (for the Computer Science Network), director Robert Kahn ensured that subscribers to the service "would not be charged for traffic." This set a new tone for openness and allowed labs and individuals with limited funds to connect over existing telephone systems.[17]

Back to a Center: Netflix and the Question of an Open Internet

Today we speak of an open Internet, one that does not discriminate against the user that produces or consumes content online. However, the meaning of "open" is shifting and needs some grounding in other, more familiar ideas that we have debated for centuries. There are two assumptions whose cultural etymologies can help us think through the issue of net neutrality and Netflix's role as an antipodal moment in the development of the Internet. The first is the assumption about equality on which U.S. society and most European countries have based policy in the past half-century. Equality, the lack of which many commentators have turned to in the past decade, is contentious in a culture where some hierarchies are thought to be legitimate. The Internet, even in its short history, has provided a destabilizing mechanism whereby anyone with minimal skill can challenge the digital architecture of even the seemingly most secure organizations. In 2014, the Internet Corporation for Assigned Names and Numbers (ICANN), the organizer of the Internet's domain name system, was hacked by employees who were tricked into opening seemingly legitimate emails full of malware.[18] Pressure from the Occupy movement as well as Massachusetts senator Elizabeth Warren's focus on consumer protectionism has forced political rhetoric to take a populist turn. Bernie Sanders's meteoric rise in the 2016 Democratic primary is clear evidence of this shift. Hillary Clinton has also had to give a nod to populism with attempts to distance herself from Wall Street, and even the Republican candidates in the 2016 U.S. Presidential election vied to be seen as the most populist candidate. The reality is that current policy directly divides income levels more than ever in the United States.

Thomas Piketty, in his authoritative study of economic inequality in the United States and Europe in the past 15 years, describes some of the reasons for the emergence of these two worlds based solely on income. There are important catalysts for their emergence that were also catalysts for the Internet's creation. In the wake of World War II, the United States and their allies were left to carve out a good part of the world as they saw fit. The world that emerged was one where power was centralized and contracts for development handed out to the highest bidder.[19] However, through its early developments through the end of the 1990s the Internet remained relatively free of government regulation. Attempts to regulate pornography and graphic images such as the Communications Decency Act of 1996 have been systematically struck down by the Supreme Court.[20] Parts of the act dealing with the production and possession of child pornography were eventually made into law. However, aside from the issue of child pornography, Congress largely avoids bills that attempt to regulate providers or users of Internet content. Congress has especially avoided *the ways in which* content is provided or distributed, which makes its recent approach to net neutrality a major shift in viewpoint and policy.

The second assumption or question that is helpful here is the question of what can truly be called "public" in the United States, and the extended question of what constitutes participation in any given public entity. The question of the Internet as a public good/right is what many see as at stake in the net neutrality case. The rise of a public good is invariably associated with the rise of the modern state. The tension between what is done in the public's name and with the public's approval is ongoing. The problem also hinges on the question of who is entitled to be a member of the public.[21] Narrowing this concept down to what constitutes a "public good" is somewhat difficult because the term is so often used inaccurately. Economist Paul A. Samuelson is credited with introducing the idea of a public good. By his description, a public good is a good whose consumption by one group does not reduce its availability to another group. Examples of public goods include fresh air, knowledge, national security, and street lighting.[22] Fresh water, as demonstrated by California's water "wars," is clearly not a public good under Samuelson's definition. Water is a utility. Public goods are frequently confused with a sense of "the public good"—something that benefits society broadly. One is a thing and the other a description. However, the water crisis in Flint, Michigan, caused by state officials knowingly allowing Flint residents to drink lead-contaminated water and leaving them without a clean water source for the foreseeable future, has started national conversation the need to see public goods and utilities as things that benefit societies broadly. As follow-up reporting on the crisis has broadened, we now know Flint, Michigan, is far from the only city with a seriously compromised water source.[23] Broad and increasing access to high-speed Internet currently fits into both of these categories. Internet access is a good whose use by one group does not affect its availability for another. For all its downfalls, most would agree the Internet has made profoundly good changes for the broader public. Here the net neutrality debate becomes somewhat more complex. The backers of a multiple-tiered Internet seek to move Internet access out of the first category, essentially arguing the Internet is nice to have but cannot continue as a good equally available to all.

The security of the public good is widely considered to be the primary goal of the state. This assumption has led to the idea in popular thought as well as in political rhetoric that the state acts on behalf of the people. The flow of information in a digitally networked society is crucial to the continuation of that society as well as the ability of its members to participate in the natural flow of networks. By the basic definition of the phrase, the Internet can be said to be a public good. Nevertheless, the networks on which our modern communications are built still produce lingering feelings of suspicion. The sense of exclusion and privilege present in the mid– to late twentieth century seem to be emerging again.[24]

Net neutrality is often presented as a seemingly straightforward issue of telecommunications companies offering one rate for broadband and dial-up connections to consumers. After five failed attempts to pass neutrality legislation in Congress, the FCC was successful in February 2015 of defining Internet access as a public utility. After silence on the issue of net neutrality for many months, FCC chairperson Tom Wheeler released a statement "aimed at preventing broadband providers from blocking, slowing down, or speeding up specific websites in exchange for payment."[25]

Not Netflix's Idea

Confined to the content of blogs and technology litigation for a decade, net neutrality became a household term when, in August 2014, Netflix began paying for more direct access to Comcast's customers—to great improvement in video streaming. The issues at the heart of net neutrality address the regulation of Internet traffic and speeds for particular customers. The term that has made Netflix synonymous with the issue of neutrality is "paid prioritization." Contrary to the way the deal has been presented by media, Netflix did not initiate this prioritization. Comcast began insisting that if Netflix was going to put so much data on its networks that it should pay for it. As Comcast spokesperson Sena Fitzmaurice said, "They choose the path the traffic takes to us. They can choose to avoid congestion or inflict it."[26] The technical results of the deal were extraordinary for Netflix. Within a week of Netflix's payments to Comcast, video quality shot up to high-definition levels. Since the deal, Netflix has begun to pay Time Warner Cable, AT&T, and Verizon for similar access to their customers.[27]

It is too easy to make Netflix the corporate villain in the narrative about net neutrality. The company has done more than many Internet-based companies to bring world class streaming content to consumers for unprecedented low prices. At the time of this essay's writing, Netflix's least expensive membership costs $9.99 per month in the United States. The streaming service offers content on a scale that was unheard of and perhaps unimaginable ten years ago when the company was shifting to streaming videos. The quality of the content has been recognized by outside organizations, with various awards for such original series as *Arrested Development* (2003–2006, 2013), *House of Cards* (2013–), and *Orange Is the New Black* (2013–) as well as the 2013 Oscar-nominated documentary *The Square*. However, it is also dangerous not to see the deal as a clearly antipodal moment in the centrifugal development of the Internet. Netflix did not seek out a tiered system or extra, better access to Comcast's customers. The fees charged to Netflix were put there initially as a penalty. However, the case has become important for the questions it raises about access, inequality, and the future architecture of the Internet. Consequently, Netflix has become simultaneously a poster child, a battleground, and a scapegoat for the issues bound up in the net neutrality debate, even after the FCC's decision.

Broadband providers have been the most aggressive opponents of net neutrality rules because the status of broadband as a public utility, while settled by the FCC, is still an open question for providers. Opponents of net neutrality make several arguments for tiered service—some stronger than others from the standpoint of precedent. One of the arguments is that since its first availability for commercial use in the late 1980s, the Internet has been a largely unregulated space. This of course is due to the engineers and developers like Baran and Crocker who envisioned a radically open space for the exchange of ideas and information. This is the actual innovation of the Internet—not the thinly veiled capital-driven innovation ISPs claim to support when they defend tiered service in the name of creativity or competition. However, the Internet is one of the few broadly public spaces where content is unregulated with the exception of child pornography. The U.S. Supreme Court consistently upholds the rights of those who post and repost videos depicting beheading and other forms of violence.[28] The notoriously right-wing group Americans for Prosperity has set up an anti-neutrality site with the ironic title "No Internet Takeover." A question remains in response to neutrality's opponents: whom do they want to prevent from taking over the Internet? However, the language that opponents and supporters of net neutrality use does not account for the unresolved, and perhaps more important, issue of what content should be prioritized.

Much of Netflix's interface, user experience, and even its corporate culture are all part of the future of video delivery and, increasingly, production. For example, CEO Reed Hastings has stated that Netflix's strategy for disrupting the way programs are made is based on prioritizing flexibility of content over efficiency.[29] The company has already dispensed with the normal series production and distribution process, instead filming and releasing a full season of episodes at once. Yet, it is unclear what the company envisions as the future. It holds the keys to that future and very publicly altered its position on paid prioritization. The populism that is in the contemporary air has caused a radical reversal in the minds of Netflix and other companies.

Nevertheless, companies have faced aggressive pushback from populist protest movements for their positions on content accessibility, as has most legislation that touches on the subject. The response to Stop Online Piracy Act (SOPA) and Protect IP Act (PIPA) are key examples of this reaction.[30] Many of the proponents of paid prioritization when it was first introduced in the Netflix/Comcast deal, such as AT&T and Comcast, have changed their stance on net neutrality, and added public statements to their websites in support of open access for all consumers. However, though the FCC ruled in favor of net neutrality in December 2015, the *Los Angeles Times* reported that Wheeler sent letters to T-Mobile, AT&T, and Comcast requesting meetings to discuss "some of the innovative things they are doing." These "innovative things" refer to data caps for streaming video. One example of a work-a-round for such caps is T-Mobile's "Binge On" initiative, which allows consumers to watch unlimited streaming video from providers like Netflix and Hulu.[31] The inflection of the language used to support or oppose net neutrality is slippery. Predictably, companies that support paid prioritization claim that neutrality legislation is detrimental to industry innovation. Meanwhile, those who support net neutrality speak of protecting consumers and keeping the Internet free and open.

Viewing the Internet as a public good creates tension with the more widely held philosophy of information-as-commodity; a philosophy through which most Internet users, operating as consumers, have worked for most of the Internet's history. The moment in which demonstrators protested the 1999 World Trade Organization's meeting in Seattle represents the first use of the Internet for widespread political purposes. One of the institutions that helped the protestors develop their online presence used the slogan "Internet for People not for Profit"—a slogan echoed by the Occupy movement a decade later. Authorities in both the United States and Canada were stunned that a sophisticated underworld of political dissidents existed online. They had not understood the history of the Internet outside of the techno-corporate relationship.[32] The Seattle moment illustrates that we can understand the controversy over net neutrality within the larger narrative of inequality that has arisen in the past decade.

Freedom and Responsibility: Netflix's Corporate Culture

Histories of the Internet tend to lionize its early engineers and interpret the development of the Internet prophetically, as the ones mentioned above do. While the founding engineers and designers were people with vision, it is difficult to say if any of them imagined a networked infrastructure on the scale that exists today. The future of the Internet is a strange thing to contemplate, and it is difficult to imagine what the online experience will be a decade or more from now. As the deal between Netflix and Comcast demonstrates, the methods, software, and quality of delivery of data shifts from month to month. The two tech bubbles in the 2000s—responsible for Silicon Valley's explosive growth and out of which Netflix was created—are evidence of the digital development that can happen in a very short amount of time. In addition, other parts of the infrastructure of the Internet change from second to second. Simply contemplating the amount of data that has been uploaded, downloaded, and networked on the Internet in the past 30 years requires comparison to amounts on smaller scales since the actual amount is difficult to conceive. However, there are organizations and initiatives that do try to predict trends for the future of the Internet and then create policy that keeps that future open for broader societal benefits. The European Commission's *Digital Agenda for Europe: A Europe 2020 Initiative* imagines a future Internet that acts as a social good but that is created by policy initiatives now and takes into account the concerns of stakeholders including individual users. Theirs is a bottom up strategy for development. They state, "The project is supported by *Futurium*, an online platform combining the informal character of social networks with the methodological approach of foresights to engage stakeholders in the co-creation of the futures that they want."[33]

Other projects and centers include the Center for the Digital Future at the Annenberg School in the University of Southern California, whose work is overseen "by a board of governors composed of the CEOs of major media and marketing companies."[34] Whether their initiatives match their longitudinal rhetoric is questionable. Harvard's Digital Futures Consortium is constantly on the lookout for what they call "truly field-changing applications of technology."[35] At a time when the word "innovative" is thrown around by universities, labs, and centers across the world, it is helpful to have a think tank to separate the truly innovative technologies and not copycats. There are other centers run by organizations such as the Pew Research Center, Elon University, and the National Science Foundation. However, there is trend with these centers: they are all at elite institutions with technically trained professionals and academics. As the events of Occupy Wall Street and the Arab Spring showed us, those who are creating the future of the Internet— or at least those that drive that future through the use of technology in the real world—are seeking solutions to social conditions. Many in Lahore and then Cairo were taken aback to hear their revolutions called "Facebook revolutions." Malcolm Gladwell of *The New Yorker* made the point, for which some at CNN criticized him, "People protested and brought down governments before Facebook was invented…. How they choose to do it is less interesting in the end, than why they were driven to do it in the first place."[36] The applications of the Internet used by Occupy and the Arab Spring might be the most innovative of all. The technologies employed—primarily social media—were not necessarily disruptive at the time they were put to use, but the ideas and willingness to participate in shared governance were what gave both movements lasting significance within the global imaginings of what the future of the Internet will be, or as Gladwell would most likely argue, what the Internet will *do*.

An interesting emphasis has emerged in the literature of all these projects. They all emphasize the importance of longitudinal feedback and user-driven policy. The future of Internet is one of the few areas of policy that governments and think tanks seek broad, popular consensus regarding its development. One may argue that this is because so many people use it for so many parts of their lives. However, this can be said about a city's water system as well, and cities rarely seek input on those systems. A more thorough explanation of the interest in popular consensus takes into account two trends. First, the Internet is a hot commodity for economic development in every sector. It has a "cool" factor, and businesses and corporations compete by outdoing one another on the digital playing field. The emergence and saturation of the market with social media marketing startups is clear evidence of businesses attempting to compete on platforms with which they are unfamiliar, yet on which they know they must be present. The second reason for the popular consensus is the careful crafting of the developers of the Internet to present and structure it as a public good or utility that, unlike other utilities like water, should remain free and open. Netflix is one of the first test cases of whether the realities of social inequality, multiplied exponentially in the United States since the 1980s, will spill over into the digital realm.

Netflix's corporate culture gives us some evidence of the intersection of the social with the digital within the ethos of Silicon Valley business culture. A dichotomous mix of rhetoric and events that nod to openness characterizes Netflix's culture, but such rhetoric does not seem to fit the reality of the kind of Internet the company imagines, or at least the Internet the company imagined before the public outcry over the deal. In the buildup to the FCC case on net neutrality, Hastings called for stricter regulation ensuring an open Internet. The buzzwords Netflix uses to describe its corporate culture are "freedom and responsibility." In a Slideshares presentation compiled by Hastings, these words appear above a yin and yang.[37] Netflix has put a lot of effort, as many Silicon Valley companies have, into defining their corporate culture against the traditional cutthroat cultures of other Fortune 500 companies.

Two other initiatives that have grown out of the idea of freedom within the corporate culture are the Netflix Prize and Netflix Hack Day. The Netflix Prize was an open competition held in 2009 for the best collaborative filtering mechanism that would beat Netflix's own algorithm for suggesting titles to users. The winning team was granted one million dollars.[38] The company regularly holds Netflix Hack Day (as many Silicon Valley companies do), and it is open to anyone, including employees, who have 24 hours to create hacks that Netflix may put into use on the site. According to Netflix's own tech blog, "Hack Day is a way for our product development staff to get away from everyday work, to have fun, experiment, collaborate, and be creative."[39] Some of the hacks have generated more lighthearted solutions, such as the technological wrinkle where partners must enter a PIN in order to watch a series together so that neither moves ahead of the other in their binge-viewing of the latest Netflix original. Others seem to have the potential to significantly improve the Netflix user interface, such as "Smart Channels" which morph Netflix browsing into an experience similar to traditional TV surfing.[40]

The hacker culture Netflix actively encourages may be one of the only checks to such imbalance in Internet access—access that the United Nations has repeatedly called a human right.[41] As mentioned above, ICANN announced that it has been hacked in the past. What do these high profile hackings mean in terms of a future Internet that works for all and not just a few well-connected users? Will the very structure of the Internet have to be rebuilt to sustain such unequal access? Internet vigilantism has created many problems for individuals as well as companies, but the hacker ethic—with origins in the very engineers who built the Internet—may hold a check for companies or governments that would abuse the massive amounts of data available to them through simple collection mechanisms.[42] In 2009, the Institute for Homeland Security Solutions at Duke University stated an obvious point and then provided a telling warning for what they believe to be the real threat of social mobilization through the Internet. The report stated, "The Internet is enabling groups previously incapable of political action to find their voices.... [This] may be relevant to understanding the potential role of the Internet for radicalization."[43]

The concern for potential radicalism, a potential built into the very protocols of the Internet by its architects, pulls back another layer and allows us to see beyond the surface debate over the deal. Perhaps the underlying basis of the national conversation about the Netflix/Comcast deal is not about net neutrality. CNET reporter Marguerite Reardon suggests that we are discussing the deal between the two companies with the wrong language: "The dispute between Netflix and Comcast is not a Net Neutrality issue because it does not have to do with how Comcast is treating Netflix's traffic once it's on the Comcast broadband network."[44] Instead, "it stems from a business dispute the two companies have over how Netflix is connecting to Comcast's network."[45] Yet even the FCC has framed the debate within the language of net neutrality. In February 2015, the FCC voted on the Open Internet Order, which classifies broadband as a utility no different from electricity or running water. Netflix, perhaps in a brilliant public relations move, was also responsible for framing the deal within net neutrality language. Reardon, who admits the deal between the two companies as well as the public debate is extremely complex, breaks down the events that led up to the debate being framed in such language:

> As anyone who has streamed video knows, the amount of bandwidth that is needed to stream or download video far outweighs the amount of bandwidth that is needed to request such a video. And the result is a massive imbalance of traffic going onto the broadband network, which likely requires a commercial interconnection arrangement between Netflix and the various broadband network providers.[46]

Netflix clearly views video streaming quality as valuable to customers, so providers should not charge extra for it. A few smaller providers have agreed to Netflix's terms. However, the major providers, Comcast and Verizon, insist Netflix pay for the interconnection. Hastings, in blog post in March 2014, made the (arguably unjustified) connection between the dispute and the issue of net neutrality. He argues, "The essence of net neutrality is that ISPs such as AT&T and Comcast don't restrict, influence, or otherwise meddle with the choices consumers make. The traditional form of net neutrality which was recently overturned by a Verizon lawsuit is important, but insufficient."[47] In this post, Hastings makes an assertion that intentionally confuses Netflix with its users. Hastings's argument is unsettling for proponents of an open Internet, because the deal prioritizes Netflix's content. By publically promoting net neutrality and opposing Comcast's insistence that companies pay for more interconnection *while also* greatly benefiting from a prioritization deal with Comcast, Hastings puts Netflix in an uncertain position. On one hand, Netflix becomes a significant public proponent of the open Internet. On the other, it has a significant amount to gain from such a landscape.

Conclusion: Inventing the Future

We can be sure the telecom industry will challenge the FCC ruling, so the debate is far from over. While the Open Internet Order was not a direct ruling on the Netflix/Comcast deal, the deal and public outrage over it was indisputably the impetus for the FCC to take another look at how big business wants to shape the Internet of the future. It is significant that major players like Netflix, whose service accounts for 30 percent of all Internet traffic, have weighed in on the side of net neutrality.[48] The company helped shape the debate and national conversation on the topic. The question remains whether or not the public wants a company, through the influence of its high-profile voice, to decide the fate of an essential public utility.

As the Internet and our ways of connecting to it and experiencing it shifts on a weekly basis, we wait to see what the FCC and companies like Netflix, Comcast, and Verizon will decide about this (now) public utility. For the moment, the experts in the centers mentioned above share a striking consensus about specific changes that will occur on/in/for the Internet in the next ten years. For example, information sharing will become seamless across devices as well as through machine intermediaries, augmented reality and wearable devices will proliferate, an "Ubernet" will emerge that will erase the meaning of national borders, and the Internet will become the "Internets" because networks will evolve and section off as the result of deals like the one between Netflix and Comcast. Yet while consensus exists on how the Internet will develop, there is much disagreement on the implications of its future. In a report titled *Digital Life at 2025*, a joint project between the Pew

Research Center and Elon University, those involved saw emerging positive relationships among societies, improved personal health, and peaceful change through publically organized protests like the Arab Spring. Yet within the same group of researchers, concerns were raised over "interpersonal ethics, surveillance, terror and crime, and the inevitable backlash as governments and industry try to adjust."[49] There was consensus amongst the researchers on two impacts of the future of the Internet. The report found that "dangerous divides between haves and have-nots may expand, resulting in resentment and possible violence" and that "pressured by these changes, governments and corporations will try to assert power—and at times succeed—as they invoke security and cultural norms."[50] The last prediction that the report makes is more of a piece of advice or a warning. The report closes, "Foresight and accurate predictions can make a difference; the best way to predict the future is to invent it."[51]

Economists have rightfully noted the Internet's role in increasing income inequality. Aside from the actual number of people in the United States who do not even use the Internet—some 60 million—there are other factors contributing to Internet-derived inequality. Low-income areas have the most to gain through increased access to the information on the Internet, yet they have been found to have the slowest connection speeds.[52] There are clear questions that governments must ask about their rural and low-income areas with slow connection speeds or none at all. If corporations have little incentive to service these areas, should governments provide the connection? If the government decides not to offer this service, are they leaving its citizens in involuntary ignorance? The Netflix/Comcast deal helps us ask these questions in a different light. Certainly, there is a digital divide—that gap between those who have access and those who do not have access to the Internet that is primarily based on income. However, what about the inequalities that exist between people who have access to the Internet? There are two additional and important ways to examine the digital divide and its impact on net neutrality and broader social inequality. First, the digital divide has grown more complex in that inequalities are deepened amongst those who have access to the Internet yet do not have the resources to interpret the information they find there.[53] The second way in which the digital divide has moved beyond access, often referred to as the "production gap" or the "second-level" digital divide, is a gap that exists between the consumers of content online and the producers of that content. The majority user-generated content on the Internet is produced by a very small percentage of the total users of the Internet. Some Web 2.0 technologies—Facebook, Twitter, YouTube—give users the ability to produce content online "without having to understand how the technology actually works."[54] This leads to an even greater divide between those who have the knowledge to interact fully with the technology and those who are "passive consumers of it."[55] The benefits of bridging the gap are clear, and little controversy exists over the fact that bridging would increase economic equality, social mobility, democracy, and economic growth.[56] Netflix has given a nod to building such a bridge, says Dennis Keseris, an intellectual property attorney. Keseris points out that Netflix and other "game-changing companies like Facebook, Google and Twitter, all of which are vocal supporters of net neutrality, have all emerged and thrived in part because of their ability to use the Internet freely, and tomorrow's web-based game-changers will need to be able to do the same in the years to come."[57]

To invent the future is a daunting task, and communities with few resources know they have more challenges in doing so or even imagining how to begin doing so. Communities are not entities that exist and then happen to communicate within themselves. They are constituted by their forms of communication.[58] Digital communities free from the oversight of a moderating center have existed for years thanks to the openness of the Internet. The promise of the digital age as envisioned by Baran and those early engineers of a radically different mode of communication was that people who wanted to could live differently—"more co-operatively and less competitively or hierarchally."[59] If counter-political movements in the past decade have taught us anything, it is that alternatives are possible—alternatives that grow out of the very decentralized networks the Internet was built on. Therefore, in this case the most productive action we can take toward keeping the Internet open is to do nothing to it at all—sort of.

NOTES

1. There is an inherent challenge in writing essays about issues that are still under debate that the writing may come across as ripped-from-the-headlines. While the legal and popular discussion of the Netflix/Comcast deal is ongoing, this essay is meant to contextualize the debate within the context of the development of the Internet as well as larger discussions about Netflix's current and potential role in maintaining an open Internet.

2. Johnny Ryan, *A History of the Internet and the Digital Future* (London: Reaktion Books, 2010), 7.

3. Jeff Sommer, "Defending the Open Internet," *New York Times*, May 10 2014, BU1.

4. *Ibid.*

5. Lily Hay Newman, "Mark Zuckerberg: 'It's Not Sustainable to Offer the Whole Internet for Free,'" *Slate*, May 4, 2015, accessed May 4, 2015, http://www.slate.com/blogs/future_tense/2015/05/04/zuckerberg_announces_changes_to_internet_org_responding_to_net_neutrality.html.

6. The Sunlight Foundation, a nonpartisan non-profit that has tasked itself with holding the government and corporations that control data more accountable and transparent, has developed a list of thirty-one policy recommendations for more transparency in data sharing and management by government agencies. See Sunlight Foundation, "Open Data Policy Guidelines," Sunlight Foundation, accessed January 3, 2016, http://sunlightfoundation.com/opendataguidelines/.

7. Physics Department University of Virginia, "Centrifugal Force," University of Virginia Physics Show, accessed January 2, 2016, http://phun.physics.virginia.edu/topics/centrifugal.html.

8. Ryan, *A History of the Internet*, 8.

9. *Ibid.*

10. *Ibid.*

11. *Ibid.*, 14.

12. *Ibid.*

13. Keenan Mayo and Peter Newcomb, "How the Web Was Won," *Vanity Fair*, July 2008, accessed January 2, 2016, http://www.vanityfair.com/news/2008/07/internet200807.

14. Paul Baran, *On Distributed Communication Networks* (Santa Monica, CA, 1962), 40.

15. Ryan, *A History of the Internet*, 32.

16. Steve Crocker, "How the Internet Got Its Rules," *New York Times*, April 6, 2009, A29.

17. Ryan, *A History of the Internet*, 92.

18. Lily Hay Newman, "ICANN Got Hacked," *Slate*, December 18, 2014, accessed December 19, 2014, http://www.slate.com/blogs/future_tense/2014/12/18/icann_hacked_in_spear_phishing_campaign.html.

19. Thomas Piketty, *Capital in the Twenty-First Century* (Cambridge: Harvard University Press, 2014), 209.

20. "Protection for private blocking and screening of offensive material," Cornell University Law School, accessed December 12, 2014, http://www.law.cornell.edu/uscode/text/47/230.

21. Craig Calhoun, "Public," *New Keywords: A Revised Vocabulary of Culture and Society*, ed. Tony Bennett, Lawrence Grossberg, and Meaghan Morris (Malden, MA: Blackwell Publishing, 2005), 282.

22. William D. Nordhaus, "Paul Samuelson and Global Public Goods: A Commemorative Essay for Paul Samuelson," speech, Yale University, New Haven, CT, May 5 2005, 2.

23. Michael Wines and John Schwartz, "Unsafe Lead Levels in Tap Water Not Limited to Flint," *New York Times*, February 8, 2016, accessed March 10, 2016, http://www.nytimes.com/2016/02/09/us/regulatory-gaps-leave-unsafe-lead-levels-in-water-nationwide.html?_r=0.

24. Frank Webster, "Network," *New Keywords: A Revised Vocabulary of Culture and Society*, ed. Tony Bennett, Lawrence Grossberg, and Meaghan Morris (Malden, MA: Blackwell Publishing, 2005), 241.

25. Gautham Nagesh, FCC Net Neutrality Plan Draws Fire from Within," *Wall Street Journal*, February 10, 2015, accessed January 3, 2016, http://www.wsj.com/articles/fcc-net-neutrality-plan-draws-fire-from-within-1423610580.

26. Zachary Seward, "The Inside Story of How Netflix Came to Pay Comcast for Internet Traffic," *Quartz*, August 27, 2014, accessed December 1, 2014. http://qz.com/256586/the-inside-story-of-how-netflix-came-to-pay-comcast-for-Internet-traffic/.

27. *Ibid.*

28. Warren Richey, "Supreme Court to Decide Case on Animal Cruelty and Free Speech," *Christian Science Monitor*, October 5, 2009.

29. "How Netflix Is Changing the TV Industry," *Investopedia*, November 3, 2015, accessed January 3, 2016, http://www.investopedia.com/articles/investing/060815/how-netflix-changing-tv-industry.asp.

30. Larry Magid, "What Are SOPA and PIPA and Why All the Fuss?" *Forbes*, January 18, 2012, accessed December 26, 2014, http://www.forbes.com/sites/larrymagid/2012/01/18/what-are-sopa-and-pipa-and-why-all-the-fuss/.

31. Jim Puzzanghera, "FCC Asking if Free-Data Plans from T-Mobile, AT&T, and Comcast Break Internet Rules," *Los Angeles Times*, December 17, 2015, accessed January 2, 2016, http://www.latimes.com/business/la-fi-fcc-tmobile-free-video-20151217-story.html.

32. Greg Elmer, *Critical Perspectives on the Internet* (Lanham, MD: Rowan & Littlefield, 2002), 28.

33. Digital Agenda for Europe, European Commission, accessed December 1, 2014.

34. Center for the Digital Future, USC Annenberg School for Communication and Journalism, accessed December 1, 2014, http://www.digitalcenter.org/.

35. HarvardX, "Digital Futures Consortium Meeting," Harvard University, accessed March 10, 2016, http://harvardx.harvard.edu/event/digital-futures-consortium-meeting-1.

36. Malcolm Gladwell, "Does Egypt Need Twitter?" *The New Yorker*, February 2, 2011, accessed December 1, 2014. http://www.newyorker.com/news/news-desk/does-egypt-need-twitter.

37. Reed Hastings, "Culture," *Slideshare*, accessed December 1, 2014, http://www.slideshare.net/reed2001/culture-1798664.

38. "Netflix Prize," accessed December 1, 2014. http://www.netflixprize.com/.

39. The Netflix Tech Blog, "Netflix Hack Day—Autumn 2015," November 9, 2015, accessed January 2, 2016, http://techblog.netflix.com/2015/11/netflix-hack-day-autumn-2015.html.

40. Adam Epstein, "The Ideas for Improving Netflix's UI were all invented by Netflix," *Quartz*, November 12, 2015, accessed January 2, 2016, http://qz.com/548058/the-best-ideas-for-improving-netflixs-ui-were-all-invented-by-netflix/.

41. Alex Fitzpatrick, "Internet Access Is a Human Right, Says United Nations," *Mashable*, July 6, 2012, accessed May 24, 2015, http://mashable.com/2012/07/06/internet-human-right/.

42. Ryan, *A History of the Internet*, 33.

43. "How Political and Social Movements Form on the Internet and How They Change Over Time," Institute for Homeland Security Solutions, November 2009, accessed December 1, 2014, http://sites.duke.edu/ihss/files/2011/12/IRW-Literature-Reviews-Political-and-Social-Movements.pdf.

44. Marguerite Reardon, "Comcast vs. Netflix: Is This Really About Net Neutrality?" *CNET*, May 15, 2014, accessed May 24, 2015, http://www.cnet.com/news/comcast-vs-netflix-is-this-really-about-net-neutrality/.

45. *Ibid.*

46. *Ibid.*

47. Reed Hastings, "Internet Tolls and the Case for Strong Net Neutrality," *Netflix*, March 20, 2014, accessed May 24, 2015, http://blog.netflix.com/2014/03/internet-tolls-and-case-for-strong-net.html.

48. *Ibid.*

49. Bridgett Shrivell, "15 Predictions for the Future of the Internet," *PBS Newshour*, March 11, 2014, accessed December 1, 2014, http://www.pbs.org/newshour/rundown/15-predictions-future-Internet/.

50. *Ibid.*

51. *Ibid.*

52. John Dilley, "Internet and Inequality: The Digital Divide Gets Personal," *betanews*, October 8, 2014, accessed December 1, 2014, http://betanews.com/2014/10/08/Internet-and-inequality-the-digital-divide-gets-personal/.

53. Mark Graham, "The Machines and Virtual Portals: The Spatialities of the Digital Divide," *Progress in Development Studies* 11.3 (2011): 211–227.

54. Colleen A. Reilly, "Teaching Wikipedia as Mirrored Technology," *First Monday* 16.1–3 (2011), http://www.firstmonday.org/ojs/index.php/fm/article/view/2824.

55. *Ibid.*

56. Digital Divide, ICT Information Communications Technology—50x15 Initiative, March 21 2014, accessed April 13, 2014, http://www.Internetworldstats.com/links10.htm.

57. Denis Keseris, "Net Neutrality: The Struggle for the Future of the Internet Has Only Just Begun," *The Telegraph*, May 23, 2015, accessed May 24, 2015, http://www.telegraph.co.uk/technology/news/11624917/Net-Neutrality-the-struggle-for-the-future-of-the-Internet-has-only-just-begun.html.

58. David Morley, "Communication," *New Keywords: A Revised Vocabulary of Culture and Society*, ed. Tony Bennett, Lawrence Grossberg, and Meaghan Morris (Malden, MA: Blackwell Publishing, 2005), 50.

59. Richard Johnson, "Alternative," *New Keywords: A Revised Vocabulary of Culture and Society*, ed. Tony Bennett, Lawrence Grossberg, and Meaghan Morris (Malden, MA: Blackwell Publishing, 2005), 4.

Doing Time
Queer Temporalities
and Orange Is the New Black

MARIA SAN FILIPPO

> Part of what has made queerness compelling as a form of
> self-description in the past decade or so has to do with the
> way it has the potential to open up new life narratives and
> alternative relations to time and space.—J. Halberstam[1]

> I'm scared that I'm not myself in here, and I'm scared that
> I am.—Piper, "Bora Bora Bora" (2013)

Breaking Out of the Primetime Prison

The two-minute, 40-second trailer used to promote the first season of
Orange Is the New Black (2013–; subsequently referenced as *OITNB*) explicitly
voices the word "time" no less than five times and invokes time's passage
throughout. The first utterance comes when WASP-y Smith College graduate
Piper Chapman (Taylor Schilling) and her Jewish writer fiancée Larry Bloom
(Jason Biggs) break the news to Piper's family about her criminal past and
encroaching incarceration on drug-related charges. Swayed by her 22-year-
old self's infatuation with, as Larry describes, "her lesbian lover who ran an
international drug smuggling ring," Piper admits to having "carried a suitcase
full of drug money once, ten years ago." In the first of her many inappropriate
responses throughout the series, Piper's haughty mother Carol Chapman
(Deborah Rush) responds, aghast, "You were a lesbian?!" "*At the time*," Piper
stresses. The scene from which this dialogue is taken appears early into
OITNB's pilot episode ("I Wasn't Ready" [2013]) and immediately signals the
conflation of criminality with lesbianism that will be voiced throughout the
series by characters such as Piper's mother who represent normative values.

75

Though omitted from the trailer, the remainder of this exchange has Piper questioned further, this time by her brother Cal (Michael Chernus), who asks, "Are you still a lesbian?" Piper replies, firmly, "No, I'm not still a lesbian." This prompts still another, only half-sarcastic, rejoinder from husband Larry: "Are you sure?" Tellingly, we do not see Piper reply; her sexuality, as we will see, is subject to logic irreducible to a single moment or label. Piper's grand-mother Celeste (Mary Looram) chimes in conspiratorially, "I once kissed Mary Straley when I was at Miss Porter's School. It wasn't for me." This cheeky exchange also announces the crucial relationship that binds both criminality and sexuality to time, wherein Piper's imminent detention stems from a long-ago infraction (she was charged with the crime two years shy of the statute of limitations expiring), while her self-identity as lesbian remains similarly rooted in another time—a sexual past that, like her criminal past, re-emerges in the present. Their combined re-emergence will have resounding effects on Piper's future, foretold by two additional lines spoken in the season one trailer: the first, issued by the formidable Russian-born inmate known as Red (Kate Mulgrew), gives voice to the terrifying possibility that Piper cannot herself acknowledge: "You'll leave [prison] in a body bag." The second line, spoken through a cell phone, has Larry responding to the news that his now-imprisoned fiancée has reconciled with former partner (and partner-in-crime) Alex Vause (Laura Prepon) by saying, "I think I need some time." Again, the last bit of the exchange is omitted, when Larry clarifies that what he needs is "time away from you [Piper]." As an upper middle class, 30-some-thing white American woman engaged to be married, Piper's two seemingly assured futures—of a long life and of marital union—are abruptly cast into doubt.

This shattering breakdown in Piper's secure recognition of herself as a "straight" citizen, in both the legal and the heteronormative senses, is signaled by the trailer's audio track sounding the plaintive opening lyrics to "A Better Son/Daughter," musician Jenny Lewis's paean to bipolar disorder: "Sometimes in the morning, I am petrified and can't move.... And [I] hope someone will save me this time."[2] Lewis's plea for salvation stands in for that of Piper, whose entrapment initially appears merely physical—15 months' confinement within the fictional Litchfield Federal Penitentiary in upstate New York—but soon reveals itself to be psychical as well, both for the toll prison takes on her men-tal well-being as well as for the mindset of white normativity that encloses her. In using Piper—who fits conventional norms of characterization (i.e. white, young, attractive, and upwardly mobile) for a "relatable" television series protagonist—as a "Trojan horse," *OITNB* enfolds viewers within an underrepresented world of disenfranchised American women whose collec-tive entrapment reveals Piper's relatively lenient sentencing to be far more predicated on privilege than punishment.[3] In so doing, *OITNB* reveals how temporality is subject to a logic governed by sexuality as well as race and class, one which uses time's binds as mechanisms of discipline and punishment to delay and deny certain citizens their pursuits of happiness.

Writing in 1990, Mary Ann Doane claimed, "The major category of television is time.... [T]ime is television's basis, its principle of structuration, as well as its persistent reference."[4] Television has undergone radical ontological, structural, and discursive shifts since the time of Doane's writing, but temporality remains persistently part of the discussion on, of, and in television. Certainly "time," as *OITNB*'s season one trailer indicates, is on these characters' minds–and on the minds of the series' creator Jenji Kohan and its distributor Netflix. In addition to the freedom, also enjoyed by cable series, from FCC content restrictions, Kohan and her writing staff enjoy temporal freedom in the form of flexibility in length. Says Kohan, "We could be anywhere from 54 minutes to an hour, depending on the episode."[5] In releasing all of *OITNB*'s first season at once, Netflix delivered a binge-watched, Emmy award-winning hit that reinforced the media platform's growing success with digitally distributed original content when it gained a reported 1.3 million U.S. subscribers (plus another 1.4 million international subscribers), thereby topping HBO's American customer base, in the financial quarter of *OITNB*'s initial release.[6] *OITNB*'s second season release in July 2014 again proved potent, pushing Netflix's customer base past the 50 million mark and boosting its international users by over a million.[7]

Echoing Doane's assessments of television's pre-millennial power, Amy Villarejo notes, "Television ushers in worldwide calendarity, a general economy of social time. Television is ... *the* implantation of social time of the twentieth century."[8] In an effort to refresh its brand since its ill-advised and short-lived 2011 attempt to bifurcate its by-mail and streaming services, and to pave the way for its entry into original content programming, Netflix in 2013 issued as part of its annual "Long-Term View" mission statement a proclamation that "'the linear TV experience' with its programs offered at set times ... 'is ripe for replacement."[9] Tellingly invoking the prison-like hold over viewers that primetime scheduling long maintained, Netflix CEO Reed Hastings characterizes the system as one of "managed dissatisfaction": "a totally artificial concept" that keeps viewers waiting—for a new episode, for a new season, and for an opportunity to discuss any given show with others.[10] In touting its ability to transform reception practices and liberate viewers from temporal mandates, Netflix ostensibly is poised to dismantle what Gary Needham calls television's power to "schedule normativity" through its authority over the "temporal coordination of the nuclear family"—alongside, I would add, that of the corporate workplace, as the proverbial water cooler that invites communal discussion of television shifts to the online realm.[11] Time-shifting capabilities have challenged this normative television schedule's corralling of what Needham calls "the marginal audience: the un-familial, the singleton, the childless couple, queers" into the "marginal zone" of post–10:00 p.m. programming.[12] But to what degree has liberation from the tyranny of time-contingent viewing, cord-dependent shared consoles, and multichannel video programming distributors (MVPDs)—i.e., cable operators like Comcast, satellite carriers like DirecTV, and fiber-optic network providers like Verizon FiOS—in favor of time-shifting, personal/mobile devices, and online video distributors such as Netflix *actually* freed viewers, temporally or otherwise, from normative family and corporate values? Does Needham, writing in 2009, need to reevaluate his claim that "it still holds that television, mass medium and commercial entity, imagines that the family audience is the ideological glue that holds it together"?[13]

Binge-Viewing, Community Purging?

Traditional television "values timeliness above all, creating a hierarchy so fundamental that it resembles natural law: New is better than old, live trumps prerecorded, original episodes always beat reruns," notes Tim Wu, whereas "online, people are far more loyal to their interests and obsessions than an externally imposed schedule."[14] Catering to trends in contemporary viewer consumption by making series available all at once places Netflix at the vanguard of industry developments in on-demand distribution. Granted unsurpassed time-shifting capability, *OITNB* fans figuratively "queered" reception practices in temporally contingent ways; for example, fans' penchant for posting episodic recaps and reflections gave way to season-long synthesizing of a sort, Kohan suggests, that is more akin to a book club than a water cooler.[15] But does the fan loyalty bred by untrammeled access translate to freedom from the temporal and non-temporal dictates of television's programming and profiteering?

While their worldwide subscription figures are a matter of public record, Netflix refuses to divulge viewing figures. Yet media analyst Procera Networks estimates, on the basis of evaluating several broadband networks, that 2 percent of U.S. subscribers, or 660,000 people, binge-watched all 13 episodes of *OITNB* in the first weekend of its season one release.[16] Despite the rhetoric of freedom with which Netflix (and time-shifted viewing generally) promotes itself, binge-viewing may operate not to enhance control but to reduce it; as Vernon Shetley notes, "The binge-watching viewer seeks instead an immersive experience, one in which, paradoxically, he or she is not in control, as the language of addiction so frequently mobilized around obsessive viewing indicates."[17] As foreshadowed by the teaser trailer for season two—which sets a quick succession of images to the insistent sound of a ticking clock and culminates in a voiceover proclaiming "3 … 2 … 1 … Go!"—Netflix acts as the pusher enabling binge-viewing with its 15-second countdown that encourages viewers to passively segue into the next episode. Inevitably, these aggressive appeals to binge have incited a backlash against its supposed ill effects, prompting Netflix to issue a series of tongue-in-cheek public service announcements featuring *House of Cards* (2013–) and *OITNB* stars' entreaties against excessive viewing.

With this shift away from the glance-flow model, television in the age of Netflix appears to depart from Michele Aaron's positing that "television with its often more distracted, channel hopping, glancing, grazing viewer would seem to depend upon a non-monogamy of viewing [that] is potentially queer."[18] Yet, once the Netflix countdown commences, perhaps other queer potentials develop in place of the non-monogamous "cruising" that Aaron as well as Jaap Kooijman associate with remote-controlled viewership.[19] Though more gazing than glancing, *OITNB*'s sofa spectatorship is conducted in home environments be they a hybrid of televisual/theatrical (using Roku boxes or other devices used to project content to a television screen) or personal/mobile (viewed on laptops or even smaller screens). This both figurative and literal queering of the space of television viewing provokes, Aaron proclaims, the "demystification of the home as haven, as homogenised, private space of (de-sexualised) hetero-romance."[20]

The pressing question regarding *OITNB*'s queer spectatorship asks whether binge-viewing serves to delimit community-formation to a near-exclusively virtual realm, a sign of the times but also a potential inhibitor to embodied and intensely affective viewership and the subcultural identity constructions that result. Unlike *House of Cards* or *Unbreakable Kimmy Schmidt* (2015–), Netflix's other early successes with original content, *OITNB* had a pre-constituted and literally queer audience eagerly awaiting its July 2013 debut, primed by advance awareness of the Piper-Alex bad romance and the casting of two lesbian cult figures, comedian Lea DeLaria as diesel-dyke inmate Big Boo Brown, and *But I'm a Cheerleader* (1999) actress Natasha Lyonne as lesbian ex-junkie Nicole "Nicky" Nichols.[21] *OITNB*'s first season would prove the tentpole for what would go on to be touted as "lesbians are having the best summer ever on TV."[22] But as Sasha T. Goldberg notes, its release amidst Gay Pride season drove the binge-viewing "gaze" and "gays" indoors for what notably remained a solitary activity rather than the communal viewing parties that constituted queer TV of not so very long ago (think of *The L Word* [2004–2009], the American version of *Queer as Folk* [2000–2005], the *If These Walls Could Talk* installments [1996, 2000], and *Ellen*'s coming out episode [1997]).[23] Amy Villarejo raises the additional question of how time-shifting disrupts queer potential when she points to a "special double episode" of *All in the Family* (1971–1979) in which Edith

(Jean Stapleton) reacts to female impersonator Beverly La Salle's brutal murder:

> The time slot for both halves of "Edith's Crisis of Faith" massively recalibrates the episode's affective stakes: they aired on Christmas night in 1977, synchronizing the episode's time (with Edith questioning her faith, as a result of homophobic violence, at Christmastime) with the time of its viewers. If the secularized and commodified ritual that is Christmas in America is here recoded as a time to devote to mourning queer loss, then *All in the Family* has achieved something significant.[24]

In contrast, the final episode of *OITNB*'s first season ("Can't Fix Crazy" [2013]) closes with a non-denominational holiday pageant staged by inmates and watched by their families and guards, which for the majority of non-diegetic viewers was incongruously Christmas in July and thus bereft of the critically queer significance that Villarejo locates.

What community-building *OITNB* generated was dispersed, mediated, and relatively anonymous, voiced in the comments sections of critics' blogs and in fans' exchanges on social media. Moreover, the binge-watching model that Netflix co-opted produces a surge in audience discourse at focused intervals—namely in the lead-up to and in the immediate wake of each season's release—that resembles the temporality of blockbusterdom's opening weekend discourse, and some have argued that the "retro" episode-per-week release structure that newly-streaming sitcom *Community* (2009–2015) selected for its second life online better serves to amplify and sustain social conversation.[25] Yet the news that *OITNB* was second only to HBO's *Game of Thrones* (2011–) as 2014's "Most Talked About TV Show" on Facebook, based on frequency of mentions in posts, would seem to discredit that notion.[26] Commenting on the choice between all-at-once and weekly release schedules, Kohan states, "It really is a double-edged sword. Part of me misses that sense of anticipation and I really miss the sense of community that you can build when everyone's watching at the same pace."[27] Certainly for non-binge-viewers and series latecomers, the threat of spoilers works as a barrier to entry into the virtual conversation. Yet online viewer communities—grouped around Twitter hashtags or fan sites, for example—are easily enough avoided by spoiler-wary users, and thus may in fact be more accommodating of the unevenness of plot knowledge than real life encounters in which the vigilance one must exercise to avoid spoilers typically operates to curb conversation, as parodied in a season three *Portlandia* (2011–) skit titled "Spoiler Alert! It's About Spoilers."

What the Netflix viewer loses in communal viewing, then, s/he seems to make up for with highly affective immersion and devotion, both in spectatorial practices of gazing, bingeing, and recapping as well as in fan-interactions aligned by shared tastes and loyalty. Phillip Maciak notes how streaming "main*streams*, to some extent, the kind of compulsive, detail-oriented mode of spectatorship we have historically associated with the *cult* or, heaven forbid, the *nerd*."[28] That association with subcultural viewers also encompasses the *queer*, and in its mainstreaming, queer spectatorship's "perverse" pleasures are de-stigmatized. Still another way in which the dynamics of virtual spectatorship are invested in the queer critique of norms is evident in the divisive debates that flourish in on-line discussions of television. Cultural critic Lili Loofbourow notes how the charged conversation that resulted from a *Game of Thrones* episode in which female character Cersei (Lena Headey) was seen by some viewers to have been raped despite the creators' insistence that the act was consensual "testifies to how robust these analytical communities have become. It shows, too, how broad a role [fictional] television has come to play in our ethical conversations."[29]

Yet the loss of embodied interactions among queer viewers in more literally subcultural sites is, without doubt, regrettable. Moreover, contemporary audiences' virtual interaction in the marketplace is cause for concern in the way it threatens to reinforce corporate profiteering in the name of expanding consumers' agency and participation. *OITNB* fans' importance for promotion of the series—and, by extension, of its corporate overseers—further fuels the millennial model of *pro bono* audience-supported media production and publicity. As with Kickstarter-funded filmmaking and older forms of web-based fan labor, Netflix exploits *OITNB*'s fervent fan base for what amounts to start-up capital *sans* fiscal return: fans create value for the corporation while receiving no financial compensation. Furthermore, Netflix's all-at-once release structure also follows a feature film model of marketing that, although advertised as giving the people what they want, takes its cue less from a participatory media model and more from the aforementioned blockbuster mentality in its reliance on demographic-targeting, saturation-booking, word-of-mouth marketing, and sink-or-swim expectations to perform.

"You've got time": The Prison as Counterpublic

Even if Netflix series do not "got time" to prove themselves before relegation to the back catalog, their narratives are anything but high-concept; whether binged-on or nibbled, their slow-build stories still pay off, albeit in the non-economic sense of telling queer histories and imagining queer futures. Thinking less figuratively about the "queerness" of viewers' time-based engagements with *OITNB*, I turn now to consider more literal and hopeful expressions of queerness in *OITNB*'s narrative transgressions of what Elizabeth Freeman terms *chrononormativity*, the institutionally and ideologically enforced temporal manipulations by which social-subjects are regulated for maximum productivity and conformity. Together with what Dana Luciano terms the *chronobiopolitical* forces that shape "'the sexual arrangement of the time of life' of entire populations," this imposed timeline functions to construct lives as productive, linear, and teleological.[30] Queer theory's deconstruction of temporality reveals time as a social construct, naturalized to those whom it privileges and in its own privileging of linearity, continuity, and progression along with such attendant chronobiopolitical discourses as progressivism, reproductive futurism, and neoliberalism. Carla Freccero views this temporal turn in queer theory as yielding "possibilities for relationality or community in queer temporal reimaginings as a way out of the repro-futurism of both hetero- and … homonormative temporal schemas."[31]

As the designated fish out of water, Piper initially reacts to the dictum "you've got time" (as sung by Regina Spektor over each episode's opening credits) with resolve to remain a productive, disciplined citizen. As she informs husband Larry about her plan for prison in "I Wasn't Ready": "I'm going to get ripped—like [fitness maven] Jackie Warner–ripped. And I'm going to read everything on my Amazon wish list. And maybe even learn a craft…. I'm going to make it count, Larry. I'm not going to throw away a year of my life." Piper's last request to her husband, upon surrendering at Litchfield, is "Please keep my website updated." Out of Piper's subsequent recognition of the assured anti-productivity that prison holds in store emerges the paradox of prison as a tool of the capitalist state that nonetheless functions in ways counterproductive to the state and its citizenry (which currently incarcerates seven million Americans at an annual cost of $74 billion) and profitable only to those with ownership stake in the prison industrial complex, the private industry our correctional system justifies and sustains. With inmates charged inflated prices for essential toiletries at commissary yet earning mere cents on the hour from their job assignments, there is no financial incentive to save or legitimate means to succeed. Some respond with extreme shows of anti-productivity and passivity (witness Nicky's waiting out her shift in the electric shop by drilling into a concrete wall an opening she claims will serve as a glory hole), some barter for supplies through underground networks such as that controlled by chef Red, and some embrace a gift economy, as with inmate Poussey's (Samira Wiley) circulating *gratis* her homemade hooch. Against these attempts to construct an alternative, less exploitative economy, two characters emerge as extreme embodiments of the capitalist ethos internalized in the role of the drug dealer: in season one, sadistic Correctional Officer (C.O.) George "Pornstache" Mendez (Pablo Schreiber); in season two, old-school inmate Vee (Lorraine Toussaint). So extreme are these characterizations that viewers point to them as the two least believable in their monstrosity, with Mendez causing a young inmate's fatal overdose and Vee staging a vicious attack on competitor Red. Other

OITNB inmates employ bartering, sharing, and non-heteroreproductive inheritance (departing inmates bequeathing certain belongings; the rest being fair game) to challenge the price gouging and labor exploiting to which they are subject. Nonetheless, *OITNB*'s prison work force is undeniably alienated, and moreover is compelled to contribute invisible, unaccountable labor commensurate with un(der)paid and un(der)regulated work performed on the outside by undocumented and domestic laborers.

OITNB makes a point of establishing that life on the outside is just as financially fraught, and its professional transactions equally exploitative. Agreeing not to borrow any additional funds from their parents, Piper and Larry are like many middle-class millennials in remaining financially dependent on their families. Though Alex (and her kingpin boss) readily turn out Piper as a drug money mule, Larry also makes the unconscionable move to sell out Piper and her fellow inmates to advance his career as a writer when he appears on a public radio talk show hosted by the wiry and bespectacled Maury Kind (Robert Stanton) in "Tall Men with Feelings" (2013). Given this narrowly disguised allusion to *This American Life* and its host Ira Glass, Larry's actions seem intended as *OITNB* creator Kohan's meta-commentary, distinguishing herself and her series from Larry's and other white males' appropriation of the stories and experiences of Others (women, people of color, queer individuals) for their own profit and for consumers' entertainment. As a silent reprimand, a voiceless and bodiless image of an unidentified young blonde woman resembling Piper, powerless in absentia, hovers over the scene of Larry and Maury recording the interview. This scene finds its converse in one that demonstrates the radically more difficult challenge put to inmates lacking Piper's cultural capital, privileged background, and support system: that of the Philip Morris–sponsored mock job fair at which inmate Taystee (Danielle Brooks) shines, but finds the rewards awaiting her nonexistent ("Looks Blue, Tastes Red" [2014]). Still more infuriatingly poignant is Taystee's actual release and return soon thereafter on a parole infraction, recidivism proving preferable to life on the outside, as she explains to prison pal Poussey in "Fool Me Once" (2013):

> When you get out, they gonna be up your ass like the KGB. Curfew every night. Piss in a cup whenever they say. You've got to go do three job interviews in a week for jobs you never gonna get. Probation officer calling every minute checkin' up. Man, at least in jail you get dinner.... Minimum wage is some kind of joke. I got part-time work at Pizza Hut and still owe the prison $900 in fees I gotta pay back. I ain't got no place to stay. Everyone I know is poor, in jail, or gone.

With its depiction of post-prison life as equally if not more fraught than doing time, and showing the two to be mutually reinforcing the ongoing disenfranchisement of America's underclass, *OITNB* unmasks what Lauren Berlant, in *Cruel Optimism*, calls "heterotopias of sovereignty": the means whereby "the good life" gets defined according to individual "freedoms" presented as a right to all but available only to the privileged few, "a fantasy that sustains liberty's normative as political idiom."[32]

As dire a description as this is of life on the outside for those abandoned by the system, it may well still seem impossible to imagine prison—the ultimate regulatory regime that dictates prisoners' movements in space and time—as a site for the disruption of chrononormativity and chronobiopolitics. The fear that *OITNB* strikes in viewers is not an echo of the famous Richard Pryor routine, "Thank God we got penitentiaries," but rather of the assessment made by *OITNB*'s avenging investigative journalist, in his appeal to Piper for inside information for his planned exposé, that our prison system constitutes "the single greatest stain on the American collective conscience since slavery" ("Comic Sans" [2014]).[33] Litchfield's officiating hierarchy features lecherous C.O.s, ruthless assistant warden Natalie Figueroa (Alysia Reiner), and most panoptical of all, her never-seen but often-invoked superior, the Big Daddy warden himself. The single gesture appearing to give inmates voice, the "Women's Advisory Committee" (WAC) for which they campaign in a season one episode, is revealed as a ruse in which those "elected" didn't actually run and are powerless to change policy ("WAC Pack" [2013]). Perhaps the most literalized symbol of prison-as-hegemony is the practice in 32 states of shackling female inmates to their maternity bed while giving birth, a threat we hear C.O. Joe Caputo (Nick Sandow) make to his subordinate officer John Bennett (Matt McGorry) to keep quiet about having impregnated Latina inmate Daya Diaz (Dascha Polanco), lest she end up "delivering [their] child with her hands and feet cuffed to the bed."[34] One also can see this power differential as simply an exaggeration of life on the outside, where pregnancy and parenting unfairly burden women, and even more unfairly those who are lower-income women of color. Painfully aware, as Daya tells Bennett in "Comic Sans," that "you have a choice, you have the power. I'm an inmate; I have nothing," she confers to him an authority already in his possession, and powerful enough to effectively undo the past, when she relinquishes him from "baby daddy" responsibility: "You got a chance right now to go back in time" ("It Was the Change" [2014]).

For J. Halberstam, "queer time and place" are constituted by temporal and spatial logics that "develop, at least in part, in opposition to the institutions of family, heterosexuality, and reproduction," and that offer alternative ways of life and modes of being that extend queerness beyond its formulation as sexual identity.[35] Considered this way, *OITNB*'s federal penitentiary setting exemplifies queer time and place: in its single-sex population; in its disruption of capitalist efficiency through its non-incentive wage system and underground economy; in inmates' resistance to time's binds—and those of the prison industrial complex—through their transformative experiences while incarcerated; and in its manifesting temporality's relation to identity by fostering sexual fluidity, with inmates who are "gay for the stay," non-monogamous, and gender variant. In all these ways, *OITNB* transforms prison into a counterpublic, which Michael Warner defines as

> publics defined by their tension with a larger public. Their participants are marked off from persons or citizens in general. Discussion within such a public is understood to contravene the rules obtaining in the world at large, being structured by alternative dispositions or protocols.... [P]articipation in such a public is one of the ways by which its members' identities are formed and transformed.[36]

We might thus consider the network of non-normative labor and kinship that *OITNB*'s inmates construct as existing outside of a capitalist and heterosexist economy, giving rise to an alternative in which, as Freeman writes, "time can be described as the potential for a domain of non-work dedicated to the production of new subject-positions and new figurations of personhood."[37]

Queer/ing Seriality and Sex

Thomas Schatz notes that for all their boundary-pushing, Netflix, like premium cable, is "staking their futures on the most traditional of television products: series programming ... signaling the persistence of the medium's most fundamental characteristics."[38] Villarejo argues that such seriality creates "density" for recurring characters that is conducive to conveying queer storylines given "that familiarity is essential to enlisting understanding and sympathy."[39] As I argue in *The B Word: Bisexuality in Contemporary Film and Television*, the temporal structure of serial narrative also crucially allows for more complex understandings of sexual identity to emerge over time.

> Where the default to status quo structure of episodic television and the contained temporality of feature films create a pressure to resolve questions of sexuality, the narrative open-endedness and expanded time-frame that characterize serial television drama offer a particularly promising site for mounting long-range and multifaceted explorations into bisexual characters' identities and experiences. Television narrative encourages bisexual representation by permitting it to unfold over time, necessary for the accumulation of experiences that renders bisexuality not practically *viable*—for any individual is potentially bisexual, no matter his or her behaviors to date—but rather representationally *legible*.[40]

When Piper clarifies in the pilot that she was a lesbian "at the time" of her drug-related crime, it is the first of many references within the series to the temporal and otherwise contingent specificity determining any utterance of identity. A flashback to Piper in the first heat of her attraction to Alex has her reporting to best friend Polly (Maria Dizzia) "I like hot girls ... and I like hot boys" ("Bora Bora Bora"). In the present tense of the same episode, a more mature Piper is less lookist but still protests "I'm not gay," when Nicky questions her about reuniting with Alex, claiming "it was about comfort, not sex," and that "I feel like I'm twenty-three and no time has passed. I think when you have a connection with someone it never really goes away." In so saying, Piper acknowledges here the importance that emotional needs and history play in determining desire, and asserts her sexuality as constantly in flux, as becoming rather than being. Characters blind to this temporal contingency of sexuality are called out for it, as in the following exchange between Larry and Piper's brother Cal from "Fool Me Once":

LARRY: Is [Piper] gay now?
CAL: I don't know about *now*. She is what she is.
LARRY: Which is what exactly?
CAL: I'm going to go ahead and guess that one of the issues here is that you think anyone needs to be "exactly" anything.

Yet despite Piper's resistance to naming herself bisexual (much-noted in the cultural discourse on the series), or ascribing to any sexuality identity in the present tense, *OITNB* effectively names her as such in characterizing her with conventional associations made between bisexuality and same-sex environment, criminality, infidelity, and white privilege. Other queer sexual identities find similarly ambivalent treatment within *OITNB*. Inmate Lorna's (Yael Stone) delusions of having a faithfully waiting fiancée on the outside and her disavowal of her carrying on with Nicky discredit her as a character, with "gay for the stay," like "hasbian" and "LUG [lesbian until graduation]," operating to contain queer sexual temporalities to the almost-extinct and the just a phase.[41] Similarly, Big Boo's dual outlaw nature as butch lesbian finds free expression in prison, whereas, Sasha T. Goldberg argues, she remains imprisoned insofar as her exceptionality within the media industry "reinforce[s] the notion of [butchness] as an outsider, solitary existence."[42] Season two reinforces this notion in painting Boo as increasingly abject, perverse, and isolated, from suggesting her carnal relationship with guide-dog "Little Boo" ("It got weird," Boo admits), to making a predatory pass at newbie Brooke Soso (Kimiko Glenn), to finally selling out Red to rival Vee and being excommunicated from both camps as a result. As I write in *The B Word*, single-sex institutions "exist both for the purpose of and as a respite from gender socialization," and screen narratives devoted to this spatio-temporal location, which include of course the women's prison film, construct a safe space for "imagin[ing] how our logics of sexual desire might be reconceived along a more fluid range."[43] While *OITNB* occasionally falls back on the women's prison tropes of predatory lesbians and mercenary bisexuals, it more prominently foregrounds romantic-erotic attachments between women. Such attachments and their capacity to flourish in all-female environments, as Freeman writes of the separate spheres of nineteenth-century Western social arrangements, are "above all temporal," operating as "havens from a heartless world and, more importantly, as sensations that moved to their own beat."[44]

Berlant and Warner use the term "sexual publics" to signify how sex and intimacy are regulated so as to maintain privacy and privatization for heterosexuality, whereas homosexuality is policed and punished according to justifications of "public decency."[45] When C.O. Sam Healy (Michael Harney), attempting to reassure a traumatized Piper during their initial consultation, says, "This isn't *Oz*," referring to HBO's prison drama that ran from 1997 to 2003, he offers a self-reflexive assurance to viewers about *OITNB* itself— in which even the rape, as defined by law, by C.O. Bennett of Daya, an inmate and thus unable to consent, is depicted in romanticized terms. Piper's own coerced instatement as "prison wife" to Suzanne "Crazy Eyes" Warren (Uzo Aduba) is defused of real threat when, in response to Piper's pal Polly asking, "Did she rape you?" Piper is forced to answer, "No, but she held my hand" ("Lesbian Request Denied" [2013]). Yet what *OITNB* decidedly does not steer away from is consensual queer sex. In the trailer referenced at essay's start, Piper is heard remarking on her love of "getting clean," reminiscing, "It's my happy place. *Was* my happy place," over images tracing an infant Piper being bathed in the sink, to an adult Piper luxuriating in the bath, to she and Larry cuddling in a tub. What the trailer omits but *OITNB*'s pilot shows in its opening moments is an additional image in which Piper and Alex, shot from the waists up and bare breasts exposed, embrace under a shower's stream. Censored from "mainstream" viewers but readily offered up to loyal subscribers at series' start, this graphic (if titillating) depiction of queer desire is something *OITNB* does not shy from showing—indeed, in another scene from the pilot, Nicky is shown with her face buried in Lorna's crotch, again while showering. What sex going on outside of prison we are shown takes place in flashback and is also queer: Alex orally pleasuring Piper on their first assignation, in early 1990s Northampton; inmate Poussey's flashback to her young adulthood in Germany, as an Army brat, getting down with a buxom fräulein. What is noteworthy is that *OITNB* looks past straight sex, ignoring such opportunities as the frenzied affair between Larry and Polly in favor of the shirtless but de-eroticized glimpse we see of her post-breast-feeding. In its most extreme instance, *OITNB* displays its predilection for queer sex and queer spaces when it sends Larry and his father to a gay bathhouse unknowingly (Mr. Bloom had a Groupon). With male frontal nudity—rarely seen on screen outside of pornography—in full line of sight, and to the strains of men copulating in corners, Larry confides: "Prison changed [Piper]. It changes people. She was not a lesbian anymore, not with me. Then she's in prison, what, a few weeks and she's a lesbian again. Or bi?

I don't even know" ("Looks Blue, Tastes Red"). As with his earlier voicing of his confusion over Piper's sexuality to Cal, here too it is Larry's lack of imagination and understanding—rather than queer sex in public—that is depicted as problematic. Because the consensual sex in *OITNB*'s Litchfield is non-phallic, subversive, and survival-minded, its occurrence constitutes moments of *jouissance* that Lee Edelman promotes as resistant to the dictates of biological procreation and capitalist re-creation encapsulated within reproductive futurism: "Detached from its reproductive function," writes Edelman, "sex can be envisioned as a subjective escape from a future pull that seems inevitable."[46] The queer sex that flourishes throughout Litchfield but tellingly sees the most action behind the altar of the under-patrolled chapel constitutes transgression of normative values both religious and repro-futuristic.

"C'mon up to the [big] house"[47]

Despite this reveling in queer sex as a non-reproductive mode of survival of the species, Litchfield's women are still forced to hear their clocks ticking within the confines of a world in which an inmate's "date" refers to her release rather than an act of romantic courtship or marriage. Aptly illustrated by the publicity image announcing *OITNB*'s second season, depicting an egg cracked in half to reveal days ticked off in anticipation, women who spend their procreative years behind bars suffer diminished opportunities to experience biological motherhood. "I told Piper it's much harder to conceive in your later 30s," Carol Chapman offers up on her first visit to Litchfield in "Lesbian Request Denied." "That's just what I need right now," Piper responds sarcastically, "a reminder of my ebbing fertility." Tried and sentenced as adults but effectively stalled in acquiring status as women (heteronormatively defined by marrying and mothering), and largely abandoned by their families once inside (if not before), inmates become wards of the state with incarceration grounds for rejection of their autonomy. Thus female inmates embody a prolonged adolescence that reveals itself as queer both in its polymorphous perversity and in its liminality—not yet contained by the perceived imperative of sexual identity labeling and its ostensible marker in sexual coupling. Given this, inmates display resistance to heteronormative and chronobiopolitical binds even as they remain bound in incarceration. Likewise, *OITNB*'s prison familial structure displays a spatio-temporality akin to the stretched-out adolescences of queer subjects and queer subcultures that Halberstam describes as "transient, extrafamilial, and oppositional modes of affiliation," while also working to reveal the alienation and perverseness that characterizes heteropatriarchal family.[48]

Biological fathers in *OITNB* are largely shown to be absent or damaging;

notably, Piper's father Bill Chapman (Bill Hoag) remains nearly as unseen as Litchfield's male warden, justifying his refusal to visit by saying in "40 Oz. of Furlough (2014)," "I'm sorry, honey, but I just can't see you like that. You're my little girl. That woman in there—that's not who you are," to which Piper responds, "That's exactly who I am." Piper's recognition of herself as inextricably tied to her past acts and present experiences constitutes a sobering yet necessary revelation in self-acceptance, signaled in the linguistic echo that links her ostensible past and present selves when she recalls, in the pilot, "It got scary, and then I ran away, and then I became the nice blonde lady I was supposed to be." With these words, Piper recounts her narrative of nonbelonging in the criminal world that she escaped to fulfill her "true" self. Yet her self-description as a "nice blonde lady" is identical to that used by Alex in convincing Piper that she can pass through airport customs, drug money in tow, without raising suspicion. In both scenarios, Piper's "nice blonde lady" is a disguise, and the white privilege it alludes to is that which allows her untrammeled access between worlds as well as the illusion of a self who is liberated from crime.

Mr. Chapman's insistence on infantilizing and idealizing his daughter finds its converse within Alex's flashback in "Fucksgiving" (2013), when her search to find her absentee birth father brings her face to face with a has-been rock star who immediately and inappropriately sexualizes her, sending her straight into the grasp of drug-dealer Fahri (Sebastian La Cause), a father surrogate under whose tutelage she will become criminalized. Nor does *OITNB* let biological mothers off the hook, but rather condemns Carol Chapman's Stepford Wife–like denials as well as the selfish mothering endured by Nicky and Daya. In parsing whether *OITNB* endorses single parenthood on the outside, however, I am skeptical; we are encouraged to raise an eyebrow when Daya's mother and fellow inmate Aleida (Elizabeth Rodriguez) comforts her about the prospect of raising alone the child she conceived with C.O. Bennett, saying, "You was raised in a non-traditional setting and you turned out great" ("Take a Break from Your Values" [2014]).[49] And though we are urged to find transgender inmate Sophia's (Laverne Cox) wife unreasonable in her plea to her then-transitioning spouse to remain anatomically male, it seems we are meant to agree with her assessment that their child needs a paternal role model when she tells Sophia, in "Lesbian Request Denied," "Do your time, so you can be a father to your son."

Berlant and Warner describe queer kinship as

> relations and narrative that are only recognized as intimate in queer culture: girl-friends, gal pals, fuckbuddies, tricks. Queer culture has learned not only how to sexualize these and other relations, but also to use them as a context for witnessing intense and personal affect while elaborating a public world of belonging and transformation. Making a queer world has required the development of the kind of

intimacy that bears no relation to domestic space, to kinship, to the couple formation, to property, or to the nation.[50]

Though under the male correctional officers' thumbs, Litchfield inmates
establish matriarchal structures that, when compassionately conceived, are
monitored both from the top down and the bottom up. Most prominently,
Red serves as the tough-but-fair mother figure first seen doling out contra-
band yogurts to her "kids," but unafraid to enact punishment when Piper's
ill-advised complaint about prison cuisine leads to Red's starving her out and
humiliating her with an English muffin-encased dish deemed "Tampon Surprise," in "Tit Punch" (2013). After Red more fervently punishes another surrogate daughter, relapsed addict Tricia (Madeline Brewer), with permanent
exile, then indirectly causes another "daughter," Gina (Abigail Savage), to be
badly scarred, Red is stripped of status as both chef and mother to her girls.
When she finally acquiesces to make amends, the reuniting of a queer family
is staged to evoke a reunion of biological family. Even accounting for the limited range of hair dyes available at commissary, the similarly red-haired, fair-
skinned women gathered around the table for Red's peace offering meal—
Nicky, Gina, Sister Ingalls (Beth Fowler), Norma (Annie Golden)—suggest
that their family ties have grown so strong as to have taken on the markers
of biological resemblance. As the apology Red proceeds to make in "40 Oz.
of Furlough" segues into a vow to chief confidante Norma, Red's words take
on the shadings of a marriage proposal: "I'm willing to make this more of a
democracy; I just want my family back…. My dear Norma, you've been by
my side for many years. You're my best friend. You've stood by me, listened
to me, plucked the hair that grows from that weird mole on the back of my
arm. I've missed you so much. Thank you for giving me another chance."
When mute Norma nods her assent, their "daughters" looking on as witnesses,
the queer family is consecrated.

It bears reminding that this commemorated queer family is one initially
formed around racial alignment; as Piper finds out upon arriving at Litchfield,
the prison population is informally organized according to a racially-ordained
kinship system that Lorna excuses as "tribal, not racist." I lack the space here
to interrogate *OITNB*'s not always enlightened treatment of race, despite its
racially inclusive casting.[51] But the troubling racial implications *OITNB* raises
in affirming the white family while pathologizing the black family are clear
enough: vindictive Vee emerges as Red's foil, functioning as the bad mother
who insists on allegiance, demands reciprocity, and reinforces cultural scripts
about dead-end prospects, shooting down Taystee's professional ambition by
chiding her, "Girl, you from this 'hood. You don't get a career, you get a job."
The perversity of the "forever family," the promise with which Vee lures the
orphaned Taystee into her criminal organization, is signaled by Vee seducing,

then ordering killed, the surrogate son who has, *sans* Vee's permission, gone into business on his own. Rematerializing out of Taystee's past to invade her present and resume her bullying, Vee exercises vigilant surveillance over the relationship between Taystee and Poussey, pressing the former, "Why do you keep defending [Poussey] over your family?" and, upon seeing them suggestively cuddling, warning Taystee, "When you get out of here you don't want people on the block talkin' about how you went that way" ("A Whole Other Hole" [2014]). Vee's homophobic surveillance of her self-appointed gang parallels C.O. Healy's, whose paranoia sends Piper to the Solitary Housing Unit (S.H.U.) when she is caught "dirty dancing" with Alex in "Fucksgiving." Vee's panoptical gaze is conjoined with that of Healy with the latter's misguided attempt to start up a therapy group he names Safe Space, wherein the quotidian threat of surveillance behind bars gets a boost from the imperative to confess. Pushing Poussey to confide her feelings, Healy's oblivious entreaty that "We have to watch out for each other" carries an irony that is anything but comforting, as only Healy is unaware that Poussey, on suspicion of squealing about Vee's heroin operation, is subject to the menacing gaze of Vee's emissary Suzanne ("Take a Break from Your Values"). Taystee eventually weans herself from the overbearing, abusive mother, telling Vee in "Hugs Can Be Deceiving" (2014), "I'm feeling like I already gave you too much of my time"; a stinging reproach, and one that *OITNB* echoes when it kills off the character whom many viewers found too villainous by half at season two's end.

Flashbacks to the Queer Future

The stylistic technique that seems most evident of Matt Zoller Seitz's observation that "in content as well as form *OITNB* is a modestly revolutionary show" is surely its use of flashbacks.[52] Despite its normalization in narrative film and television, the flashback is stylistically radical in its reflexivity and non-linearity, and its use nearly always signals disturbance—whether of a traumatic memory impinging on the present or of a recollection so pleasurable that is provokes a yearning to return to the past. Moreover, as Maureen Turim notes, flashbacks serve to interpolate viewers into identifying with characters through the subjective revealing of history:

> If flashbacks give us images of memory, the personal archives of the past, they also give us images of history, the shared and recorded past…. This process can be called the "subjective memory," which here has the double sense of the rendering of history as a subjective experience of a character in the fiction, and the formation of a Subject in history as the viewer of the film identifying with fictional character's [*sic*] positioned in a fictive social reality.[53]

Flashbacks are stylistically as well as temporally radical in being by definition non-teleological, resisting the imperative towards progress and futurity's pull and in so doing evincing the famous Faulkner axiom "The past is never dead. It's not even past."[54] In a singular instance within "Lesbian Request Denied," the past fuses with the present with the use of a single (trick) shot showing transgender inmate Sophia's transition from her assigned gender identity as "Marcus" (played by actress Cox's real life twin brother M. Lamar) to her chosen identity. Given this episode's focus on Sophia's panic when budget cuts deny her the necessary estrogen dosage, this time-shifting shot visualizes the constant threat of slippage back into the past that renders a gender-transitioning inmate dependent and desperate. Indispensable for establishing viewer identification with each inmate among *OITNB*'s ensemble, these episodic forays into characters' pasts via flashback establish the unfortunate constellation of factors that led to each woman's incarceration; overwhelmingly, it was lack of economic resources (whether for material needs, education, healthcare, or legal defense) that was significantly if not wholly to blame. But even as these individual histories reveal the extent and consequences of America's poverty epidemic, they also revise cultural scripts by resisting ahistorical, decontextualized narratives. The result is to unbind characters from the representational regime of stereotyping as well as from the legal regime of criminalizing, humanizing each character and qualifying if not exonerating her of her crime(s).

As the season one trailer discussed at essay's start indicates, in the eyes of her biological, traditional family, Piper's sexually curious susceptibility to being seduced by Alex is what ostensibly led to her incarceration and perceived failure as a straight citizen, in both senses. This conflation of queerness with failure—of heterosexuality, of repro-futurity, of bourgeois capitalism— is ideologically ingrained. "The queer body and queer social worlds become the evidence of that failure," writes Halberstam, "while heterosexuality is rooted in a logic of achievement, fulfillment, and success(ion)."[55] *OITNB*'s other "bad family" and its matriarch Vee share the elder Chapmans' conflation of non-futurity with failure, evidenced in Vee's scolding of her surrogate daughter "Black Cindy" (Adrienne C. Moore) when the latter blows off her orders as Vee's drug deputy, in "Comic Sans." "That probably worked back when folks were expected to check out at forty," lectures Vee. "Thing is, if you're not building a future, that's because you don't believe there is a future. You've given up on yourself. You're a loser." The perverse work ethic that Vee preaches, though an echo of capitalist dogma, all but guarantees a future behind bars if not the life cut short that awaits Vee. Yet neither does *OITNB* offer the blindly *carpe diem* rejection of futurity that Alex proposes to Piper, saying, "I'm not planning anything. I don't know what's going to happen, that's the point of being with me"—a seductive fantasy but one Piper (and we) have by season two's end come to understand as avoidant and self-serving.

Edelman's urging for a queer oppositionality that is future-negating did not find favor throughout queer theory.[56] Without succumbing to the hollow notion that "it gets better," queer scholars of color in particular see it as essential to maintain belief, as José Esteban Muñoz attests, that "queerness is primarily about futurity and hope."[57] For the disenfranchised characters of *OITNB* and their real life counterparts, it is difficult to deny that which Dustin Bradley Goltz posits: "Given the way discourses of future punish queerness, hope marks a site of political struggle and urgency, for, as [Gloria] Anzaldúa reminds us in *This Bridge Called My Back*, 'hopelessness is suicide.'"[58] For Goltz, this more moderate position imagines queerness contending with heteronormativity's "ongoing project of hijacking the future" by vigilantly critiquing the status quo, "working within the present—rejecting the tragic forecasting of heteronormativity's threats, with a 'day at a time' orientation towards unseen futures."[59] Above all, *OITNB*'s characters challenge what our punitive rather than rehabilitative system of incarceration dictates: rather than killing time, they queer time, producing what Freeman describes as "new social relations and even new forms of justice that counter the chrononormative and chronobiopolitical ... forms of time on which both a patriarchal generationality and a maternalized middle-class domesticity"— and, I would add, a neoliberal capitalist economy—"lean on for their meaning," with the result being one "of transformation rather than of shared victimhood."[60]

Discourses on television's futurity steer to similar extremes, with dystopian forecasts of "the end of television" countered by utopian rhetoric of creative autonomy, indie TV, virtual entrepreneurs, and participatory media. In assessing whether Internet television *really* constitutes an overhaul of the corporate management and economic arrangements profiting Hollywood's big media groups, it is important to recall that Netflix turned to original content in an attempt to compensate for losing the virtual monopoly it had acquired as an early adopter of media streaming, as expired licensing deals and rising competition from rival content providers like Hulu and Amazon increasingly threatened Netflix's market advantage. Also crucial to remember is that it is not, for the foreseeable future, within Netflix's control to radically shift the entertainment industry's power structure, given that Netflix remains dependent on licensing fees, priced at the discretion of film and television studios. Even original content such as *OITNB* is "exclusively" controlled by Netflix only for first-window streaming rights. Suggesting a still hazier future, some media analysts predict the likelihood that cable-systems operators will compensate for the loss in profit from bundling cable packages (as the result of "cord-cutters" and "cord-nevers") by shifting their business model to focus on selling broadband service, such as Verizon and AT&T are already doing.[61] In the event that these broadband providers successfully appeal the FCC's recent passing of net neutrality regulations protecting streaming services, Netflix could be faced with inflated costs or data caps for streaming bandwidth.[62] Netflix chief content officer Ted Sarandos freely admits to how its non-normative programming model is economically profitable, even as he hitches that financial incentive, in the very next sentence, to a nationalistic notion of freedom:

> To make all of America do the same thing at the same time is enormously inefficient [and] ridiculously expensive…. There is a freedom achieved when your options extend beyond the night's offerings and the limited selection of past episodes that networks make available on demand. Specifically, it's the freedom to only watch television you enjoy.[63]

In linking viewer freedoms to free market policies, Sarandos positions Netflix as the neoliberal alternative to an outmoded model redolent of no less than socialism in its implied conformity and scarcity. Yet as the first generation of Internet television viewers have discovered, the liberation that comes with expanded access and choice may prove overwhelming and ultimately impoverishing. "With Internet television, customers have grown accustomed to the idea that they might really have it all, whenever they want," writes Ken Aueletta. "What's unclear is whether that freedom will cost more than the current bargain."[64]

Though it may be seen as constituting a similarly utopian disregard for reality and mortality, *OITNB*'s second season ends with an image that evocatively visualizes queer imaginings of individual and collective transformation realizable through a radical temporality that transcends past, present, and future and the chrononormative, chronobiopolitical imperative each confers. Handed a death sentence in the form of a terminal cancer diagnosis with no prison funding for further care, former bank robber and long-time convict Rosa (Barbara Rosenblat) seizes her moment to escape, commandeering the prison transport van to speed off into the proverbial sunset to the rising chords of Blue Öyster Cult's "(Don't Fear) The Reaper" ("We Have Manners. We're Polite" [2014]). At this moment of liberation from the various incarcerations of place, class, race, age, and illness, Rosa visually transforms back into the brazen firebrand she once was. In its ecstatic affirmation of a self who is unconstrained by either time's binds or that of socially dictated subjectivity, it is a moment of untrammeled *jouissance*. Alas it is but a fantasized escape, yet in its projection on screen we can imagine its possibility. At their best, both Netflix and *OITNB* offer a similar view beyond, to a queer time and place both televisual and societal.

NOTES

This essay is expanded from and inspired by a brief visual essay I published as part of an *Orange Is the Black* theme week in *In Media Res*, and I am indebted to my co-contributors and our respondents within that forum for their valuable insights. See Maria San Filippo, "Doing Time: Queer Temporalities in *Orange Is the New Black*," *In Media Res*, March 10, 2014, accessed January 22, 2016, http://mediacommons.futureofthebook.org/imr/2014/03/10/doing-time-queer-temporalities-and-orange-new-black.

1. J. Halberstam, *In a Queer Time and Place: Transgender Bodies, Subcultural Lives* (New York: New York University Press, 2005), 2.

2. "A Better Son/Daughter," the first track on the 2002 album *The Executioner of All Things*, was recorded when Lewis was the frontwoman for the band Rilo Kiley.

3. *OITNB* creator Jenji Kohan referred to Piper as "my Trojan horse" on *Fresh Air*, *NPR*, August 13, 2013, accessed February 16, 2016, http://www.npr.org/2013/08/13/211639989/orange-creator-jenji-kohan-piper-was-my-trojan-horse.

4. Quoted in Gary Needham, "Scheduling Normativity: Television, the Family, and Queer Temporality." In *Queer TV: Theories, Histories, Politics*, ed. Glyn Davis and Gary Needham (New York: Routledge, 2009), 144.

5. Kohan, *Fresh Air*.

6. Victor Luckerson, "After Disaster, Netflix Is Back from the Brink," *Time*, October 21, 2013, accessed January 22, 2016, http://business.time.com/2013/10/21/how-netflix-came-back-from-the-brink/.

7. Sonali Basak, "'Orange Is the New Black' Vaults Netflix Past 50 Million Users," *Bloomberg*, July 21, 2014, accessed January 22, 2016, http://www.bloomberg.com/news/2014-07-21/-orange-is-new-black-vaults-netflix-past-50-million-subs.html.

8. Amy Villarejo, *Ethereal Queer: Television, Historicity, Desire* (Durham: Duke University Press, 2014), 77.

9. Ken Auletta, "Outside the Box," *The New Yorker*, February 3, 2014, 54.

10. Nancy Hass, "And the Award for the New HBO Goes To," *GQ*, February 2013, accessed January 22, 2016, http://www.gq.com/entertainment/movies-and-tv/201302/netflix-founder-reed-hastings-house-of-cards-arrested-development.

11. Needham, "Scheduling Normativity," 145.

12. *Ibid.*

13. *Ibid.*

14. Tim Wu, "Netflix's War on Mass Culture," *New Republic*, December 4, 2013, accessed January 22, 2016, http://www.newrepublic.com/article/115687/netflixs-war-mass-culture.

15. Kohan, *Fresh Air*.

16. Emma Daly, "How Many People Are Watching Netflix?" *RadioTimes*, February 27, 2014, accessed January 22, 2016, www.radiotimes.com/news/2014-02-27/how-many-people-are-watching-netflix.

17. "Comments," in San Filippo, "Doing Time."

18. Aaron is careful to clarify that this notion of queer viewership as non-monogamous "is not to associate it with promiscuity ... I am not rendering queer untrue, only uncontained, and in doing so bringing into relief the normative logic that it offsets." Michele Aaron, "Towards Queer Television Theory," in *Queer TV: Theories, Histories, Politics*, ed. Glyn Davis and Gary Needham (New York: Routledge, 2009), 71–72.

19. See Jaap Kooijman, "Cruising the Channels: the Queerness of Zapping," in *Queer TV: Theories, Histories, Politics*, ed. Glyn Davis and Gary Needham (New York: Routledge, 2009), 159–171.

20. Aaron, "Towards Queer Television Theory," 72.

21. *Unbreakable Kimmy Schmidt* was originally slated for broadcast on NBC but was sold to Netflix.

22. Margaret Lyons, "Lesbians Are Having the Best Summer Ever on TV," *Vulture*, July 25, 2013, accessed January 22, 2016, http://www.vulture.com/2013/07/lesbians-are-hav ing-the-best-summer-ever-on-tv.html.

23. "Comments," in San Filippo, "Doing Time."

24. Villarejo, *Ethereal Queer*, 90.

25. See, for example, Jim Pagels, "Stop Binge-Watching TV," *Slate*, July 9, 2012, accessed January 22, 2016, http://www.slate.com/blogs/browbeat/2012/07/09/binge_watching_tv_ why_you_need_to_stop.html.

26. Timothy Stenovec, "One of the 'Most Talked About TV Shows' of 2014 Wasn't on TV," *Huffington Post*, December 9, 2014, accessed January 22, 2016, http://www.huffington post.com/2014/12/09/orange-is-the-new-black-most-talked-about-facebook_n_6291114.html.

27. Kohan, *Fresh Air*.

28. Phillip Maciak, "Streaming Pam Beesley," *Los Angeles Review of Books*, October 9, 2013, accessed January 22, 2016, http://blog.lareviewofbooks.org/deartv/streaming-pam-beesly/; emphasis in original.

29. Lili Loofbourow, "How Recaps Changed the Way We Think about TV—and Our Lives," The *Guardian*, November 4, 2014, accessed January 22, 2016, http://www.theguardian. com/tv-and-radio/2014/nov/04/how-recaps-changed-the-way-we-think-about-tv.

30. Quoted in Elizabeth Freeman, *Time Binds: Queer Temporalities, Queer Histories* (Durham: Duke University Press, 2010), 3.

31. Carla Freccero, et al., "Theorizing Queer Temporalities: A Roundtable Discussion," *GLQ*, 13.2–3 (2007): 187.

32. Lauren Berlant, *Cruel Optimism* (Durham: Duke University Press, 2011), 98.

33. I thank Vernon Shetley for making me aware of Pryor's routine, viewable at (accessed February 16, 2016) http://www.dailymotion.com/video/xc50lv_richard-pryor-on-arizona-penitentia_shortfilms.

34. I thank Daisy Ball for bringing this issue of the shackling of pregnant prisoners to my attention; see Ball, "The Essence of a Women's Prison: Where *Orange Is the New Black* Falls Short," *In Media Res*, March 14, 2014, accessed January 22, 2016, http://mediacommons. futureofthebook.org/imr/2014/03/14/essence-womens-prison-where-orange-new-black-falls-short. For the International Human Rights Clinic's report, see https://ihrclinic.uchicago.edu/ page/shackling-pregnant-prisoners-united-states.

35. Halberstam, *In a Queer Time and Place*, 1–2.

36. Michael Warner, *Publics and Counterpublics* (New York: Zone Books, 2005), 55–57, 66.

37. Freeman, *Time Binds*, 54.

38. Thomas Schatz, "HBO and Netflix—Getting Back to the Future," *Flow* 19 (2014), http://flowtv.org/2014/01/hbo-and-netflix-%E2%80%93-getting-back-to-the-future/.

39. Villarejo, *Ethereal Queer*, 90.

40. Maria San Filippo, *The B Word: Bisexuality in Contemporary Film and Television* (Bloomington: Indiana University Press, 2013), 203–204; emphasis in original.

41. See Ben Davies and Jana Funke, "Introduction: Sexual Temporalities." In *Sex, Gender and Time in Fiction and Culture* (New York: Palgrave Macmillan, 2011), 3.

42. Sasha T. Goldberg, "'Yeah, Maybe a Lighter Butch': Outlaw Gender and Female Masculinity in *Orange Is The New Black*," *In Media Res*, March 12, 2014, accessed January 22, 2016, http://mediacommons.futureofthebook.org/imr/2014/03/12/yeah-maybe-lighter-butch-outlaw-gender-and-female-masculinity-orange-new-black.

43. San Filippo, *The B Word*, 130, 132.

44. Freeman, *Time Binds*, 5.

45. See Lauren Berlant and Michael Warner, "Sex in Public," *Critical Inquiry* 24.2 (1998): 547–566.

46. Quoted in Davies and Funke, "Introduction: Sexual Temporalities," 9. See also Lee Edelman, *No Future: Queer Theory and the Death Drive* (Durham: Duke University Press, 2004).

47. This section heading references the Tom Waits track "C'mon Up to the House" that plays over the closing moments of Piper's furlough episode, when she finds herself alone and alienated from her former life, and thus however improbably drawn back to the new normal of prison ("40 Oz. of Furlough" [2014]).

48. Halberstam, *In a Queer Time and Place*, 154.

49. Aleida gradually redeems herself in expectation of becoming a grandmother, such that Daya "has two mommies": biological mother Aleida and adoptive mother Gloria (Selenis Leyva).

50. Berlant and Warner, "Sex in Public," 558.

51. For a cogent discussion of *OITNB*'s treatment of race, see Jennifer L. Pozner, "TV Can Make America Better," *Salon*, August 29, 2013, accessed January 22, 2016, http://www.salon.com/2013/08/29/tv_can_make_america_better/.

52. Matt Zoller Seitz, "Department of Corrections," *New York*, June 2–8, 2014, 95.

53. Maureen Turim, *Flashbacks in Film: Memory and History* (New York: Routledge, 1989), 2.

54. William Faulkner, *Requiem for a Nun* (New York: Vintage, 2011), 73.

55. J. Halberstam, *The Queer Art of Failure* (Durham: Duke University Press, 2011), 94.

56. Edelman, *No Future*.

57. I am referencing the It Gets Better Project, a public service campaign founded in 2010 by Dan Savage and Terry Miller, aimed at preventing suicide among bullied LGBT youth. José Esteban Muñoz, *Cruising Utopia: The Then and There of Queer Futurity* (New York: NYU Press, 2009), 11.

58. Dustin Bradley Goltz, "It Gets Better: Queer Futures, Critical Frustrations, and Radical Potentials," *Critical Studies in Media Communication* 30.2 (2013): 139.

59. Dustin Bradley Goltz, *Queer Temporalities in Gay Male Representation: Tragedy, Normativity, and Futurity* (New York: Routledge, 2010), 151.

60. Freeman, *Time Binds*, 10, 46, 49.

61. See, for example, Auletta, "Outside the Box," 61.

62. See Marguerite Reardon, "13 Things You Need to Know about the FCC's Net Neutrality Regulation," *CNET*, March 14, 2015, accessed January 22, 2016, http://www.cnet.com/news/13-things-you-need-to-know-about-the-fccs-net-neutrality-regulation/.

63. Quoted in Wu, "Netflix's War on Mass Culture."

64. Aueletta, "Outside the Box," 61.

Netflix and Innovation in *Arrested Development*'s Narrative Construction

MAÍRA BIANCHINI
and MARIA CARMEM JACOB DE SOUZA

In "Development Arrested" (2006), *Arrested Development*'s (2003–2006) third season finale, Maeby Fünke (Alia Shawkat), then a successful teen Hollywood executive, presents the idea of a television series based on her family to an industry icon, who says, "Nah, I don't see it as a series. Maybe a movie." The icon in question was of course Ron Howard, the series' executive producer and narrator, and this moment hinted at the potential plans *Arrested Development*'s creative team had for a film sequel to the Bluth's family story.

Seven years later, in "Señoritis" (2013), a fourth season episode made available first on Netflix, it is Michael (Jason Bateman) who embarks on a journey to get the permission of his family members for a potential feature film. Michael hears from his now unemployed niece about the *new* promising plan for the project: "I gotta tell you, I think movies are dead. Maybe it's a TV show." Taken with Ron Howard's speech in the original finale, these lines of dialog make explicit reference to the real-life story of how the series was revived after its 2006 cancellation—a creative saga that includes frustrated investments in a feature film and a return as, once again, an episodic series on Netflix.[1]

Season four's online distribution makes us wonder about the reasons why Netflix's executives decided to support Howard's and executive producer Mitch Hurwitz's project. The company's actions in the recent audiovisual multiplatform market show Netflix's interest in increasing its importance in this sphere, especially, in the field of television series production, a social

98

field where media convergence and participatory culture are strong features. Ted Sarandos, Netflix's Chief Content Officer since 2000, has been acknowledged as one of Netflix's executives responsible for translating his identification of promising series production market trends into consecrated original programming strategies designed for an audience who has proven to be faithful consumers of the aesthetic and stylistic innovations fostered in this context.

The decision to bring back showrunner Hurwitz and his creative fellows for a fourth season of *Arrested Development* illustrates Netflix's expertise in combining the series' captive fan base with the creative team's interest in remixing the American sitcom form. In its original run, the series was recognized for its formal inventions, including the absence of laugh track, a single camera (as opposed to three camera) set up, "cinéma verité" documentary-style aesthetics (with handheld cameras, the presence of a narrator, and the use of captions, voiceovers, and collages), and, most importantly, its self-conscious, multi-layered, and complex mode of storytelling.[2] Thus, Netflix demonstrates how to be in tune with a series' fan community cultivated online, a connoisseur group of spectators skilled at appreciating the production's operational aesthetics and bold stylistic choices made by the anthological, radically fragmented narrative construction of the fourth season, with episodes focusing on different perspectives of the same story.[3] The structure of the season's comic effect, through temporal alternations and narrative fragmentation, was designed by exploring the possibility of continuous viewership of the 15 episodes—utilizing the binge-watching behavior associated with Netflix's distribution model. Such a decision ultimately resulted in a radicalized puzzle-like design of the narrative.

This essay examines the intricate relationship between the ideas of Netflix's executives and *Arrested Development*'s creative agents. Such ideas and decisions generated, at once, an aesthetic renovation of the series, an appreciation of the creative team, and the company strengthening its position in the scripted television marketplace. To examine such relations between the company's executives and *Arrested Development*'s creative team, we rely on Pierre Bourdieu's research, which enables a better understanding of these agents' actions through a relational analytical approach of the logic governing the disputes for distinction and recognition in the history of the television series production field.

The concept of a social field of television series production is applied here to show how television series are created according to specific rules and logical principles which enable a system of both partnership and competition relations amongst the agents—from company's managers to creative professionals working on such products—involved in the processes of original programming production and distribution. The principles that encourage these agents' actions are the result of the field's historical autonomization process, which can be observed in the disputes over power to define and legitimize a series format and quality as well as the power to evaluate, recognize, and consecrate the agents that produce them.

Therefore, in the history of the production, distribution, and consumption of television, a market of economic and symbolic exchanges is formed and governed by shared structural dimensions, homologous positions in relation with other fields (such as the dominant/dominated positions and these poles' intermediate positions found in the economic and political fields); meanwhile, private and autonomous dimensions are formed and structure the conflicting social positions over the definition of an original, innovative, and quality television series. The central premise of this analytical perspective shows how the poetic choices orchestrated by *Arrested Development*'s creator Hurwitz tend to be guided by the dispositions, interests, and positions he occupies within the television industry, both at the time the series aired on Fox and at its revival on Netflix.

Position-Taking and Autonomy in Television's Social Field

Before *Arrested Development*, Hurwitz had a career in traditional multi-camera sitcoms, most notably on *The Golden Girls* (1985–1992), and he found the format to be outdated. Hurwitz's conditions to experiment with new forms of comedy storytelling were made possible through some synergistic thinking by Hurwitz, Ron Howard, producer Brian Grazer, and Imagine Entertainment chief David Nevins in 2002. As Hurwitz said, "The idea was that this documentary style could inform the storytelling, that instead of everything being consecutive, tell the story in a non-linear way. It became stylistically this great way of moving the story along."[4]

Arrested Development's emergence in this scenario inspires analysis on the performances of the production and distribution companies in the industry. The degree of autonomy granted to series creators generally depends on the history of autonomization in the field and on the symbolic and economic capital accumulated by such creators. What this notion inspires is an examination of how those working within media industries respond to external (like the need for commercial success) and internal (a production field's specific defining criteria for quality and formal aspects) hierarchies. Following this direction, this essay employs a broader analysis of the professionals working in companies responsible for producing and distributing media content products within the concept of the social field. Entrepreneurial social agents are specially fascinating to observe, as they dedicate themselves to fostering industrial competitiveness, enhancing communication skills and marketing strategies, and bringing together experts who illustrate a combination of "economic dispositions which, in certain field's sectors, are totally alien to the producers, and intellectual dispositions similar to those of the producers, from whom they may exploit the work only to the extent they know how to appreciate and value it."[5] Such executives tend to adopt a legitimation stance regarding creative and innovative ideas.

Research on the development of American television series illustrates the creative experience that includes the collective work of producing a series, headed by a showrunner (a professional with a high degree of autonomy to manage the key areas of writing, producing, and directing), depends on the conditions offered by the companies' executives.[6] Showrunners are able to increase their degree of autonomy and control over the complex process of production to the extent that they can accumulate symbolic capital and encounter homologous positions amongst executives from broadcast networks, cable and premium channels, and, most recently, online streaming companies.[7] In this context, showrunners know the higher the power to negotiate and interfere with the process, the better the conditions to choose the resources and strategies employed. This may provide the recognition of their authoring features, which tend, in turn, to be associated with producing companies' brands. A remarkable example is HBO's original programming from the 1990s and early 2000s. Showrunners such as Tom Fontana (*Oz*, 1997–2003), David Chase (*The Sopranos*, 1999–2007), Alan Ball (*Six Feet Under*, 2001–2005), David Simon (*The Wire*, 2002–2008), and David Milch (*Deadwood*, 2004–2006) exercised the creative freedom enabled by the premium cable channel and created some of the most critically acclaimed dramas in television history, which subsequently helped elevate HBO's brand in the field of television production.[8] Executives use brand identity strategies to establish their companies, investing in original programming and selecting series that, preferably, find a convergence point between aesthetic quality and audience ratings.[9] Innovative series thus seem to demonstrate a tendency to become an object of adoration to niche audiences.

This essays considers how, in the field of television production, the increasing recognition and consecration of executives, creative writers, and producers along with specific genres and formats, come together to stimulate innovative processes in the spheres of creation, distribution, and consumption of programming. Such innovations show how the current state of the field's autonomization allows legitimized series to become aesthetics reference points in American television. Thus, key moments of the main social agents' trajectories are presented throughout the essay. Such agents include: *Arrested Development*'s creator and showrunner, Hurwitz; producers and executives of Imagine Entertainment, Ron Howard, Brian Grazer, and David Nevins, who bet on the narrative and aesthetic style of the series from its inception; and Ted Sarandos, one of the executives responsible for the fourth season on Netflix who encouraged the series to further explore its previous innovations.

The social trajectory of a showrunner, associated with the network's executives that legitimize them, tends to be a privileged *locus* of observation of the association efforts towards the creation of distinctive brands that stand out in both the serial production's work and in the wider sphere of practices conducted by the companies involved.[10] Analyzing the social consecration trajectory of these companies and entrepreneurs therefore enhances the understanding of efforts to promote their position in the markets in which they operate.

The business strategy of creating favorable and autonomous conditions for showrunners to work demonstrates that the creation context for successful television series can also be a circumstance that consecrates the company. This attitude is risky but necessary, particularly in a time of change in such fields, where the results from risky bets may positively reflect on the rise of newcomers in competitive spheres, like Netflix in the television production and distribution market.

The purpose of the reflection here is to indicate the perspective these companies advocate for regarding innovations in television sitcoms, which can be observed in the narrative strategies employed by *Arrested Development*. This essay also illustrates that the actions of the social agents involved have repercussions on the expansion of quality programming. In their success, these social agents strengthen the value and autonomy of creative and original content, and aid in the rise of new entrepreneurial agents such as Netflix in a moment of intense changes in the field.

"The future of television is here": Netflix in the Field of Television Series Production

The expansion and diversification of how television content is distributed and consumed has significantly altered the dynamics of the field of television production. Companies like Netflix, which at first only offered a new distribution platform, have begun original series production. Netflix has projected itself into a field that was, until recently, restricted to production companies, networks, and channels. When acting in this field, Netflix has employed familiar industry strategies, with HBO's strategic planning as main reference—as stated by Netflix's co-founder and CEO, Reed Hastings.[11] Such companies deal with at least three associated features: (1) international market expansion; (2) the creation of a system to regionally and globally identify, attract, assess, retain, and expand subscribers; and (3) the combination of a set of products that live up to the diversity and wishes both of its wider and its niche audiences. In such cases, the goal is innovative original content, recognized by consecration instances represented by experts in the field (critics, producers, authors), and the specific audience-related recognition (fan communities, websites, specialized blogs).

The strategies employed by Netflix demonstrate how the use of a new technology modifies the positions of the field's players. The analytical perspective developed by Bourdieu highlights the newcomers' tendency to alter the balance of forces significantly in a given historical moment, an essential aspect of the field's logic and operating dynamics. In recent years, new agents have entered the field of series production, with the significant focus on the emergence of online streaming platforms. Along with Netflix, Amazon, Hulu, and Yahoo have all invested in original programming strategies, achieving varying degrees of success and recognition. While Amazon's *Transparent* (2014–) received Golden Globe and Primetime Emmy attention, Yahoo lost over $42 million with its original content, including the revival of cult NBC sitcom *Community* (2009–2015).[12] The field's symbolical capital is in constant negotiation, favoring the ascendance of agents who better combine strategic planning and creative potential in the space of choices made available in this field of production.

The way Netflix positioned itself within the field of series production and distribution companies, achieving recognition for the quality and innovation of its series, demonstrates that we are facing a new configuration of disputes and power relations in the multi-platform ambience. Netflix displays an ability to grasp the principles of the social field's logic and to identify changes that destabilize the dynamics of existing operations—regarding technologies and content distribution systems under development, the expanding global market, the rise of new audiences, the harbor of audacious creators, and new state regulations, among others. Executives know how to evaluate this new context, and they implement action strategies that put them in a new position while competing with players who have been on the market for decades, that need, in turn, to renew and innovate their business models in light of these systems' intrinsic logic.

In Netflix's case, the investment in original programming can be understood through the history of the company's content distribution model, when it ceased to be just an online rental service and it started working as an on-demand streaming platform. Two initial challenges presented themselves to Netflix in this transition period in 2007: licensing and the restriction of content distribution windows. According to Sarandos, the absence of a first-sale doctrine to the streaming market demanded a reconfiguration in the licensing of content distribution agreements, particularly with television content.[13] The first-sale doctrine allows a company to purchase a large collection of movies and television series as physical media and distribute these products indefinitely, since the copyright holders of such products do not have a perpetual license on them. That is, while the physical integrity of DVDs is preserved, the company does not need to pay again for the right to rent or resell the product. In the streaming market, on the other hand, the companies need to ensure a subscription video-on-demand (SVOD) television right, which must be renewed constantly, in a competitive dynamic with other SVOD players. According to Sarandos, such negotiations for television content become progressively more complex, since "every network is interested in holding, withholding, buying, or blocking SVOD rights as a way to create an atmosphere for their own VOD services."[14]

One major challenge regarding distribution windowing came from established agreements among Hollywood studios and cable channels, both premium and basic. These longstanding agreements gave cable channels exclusive first rights to major releases for a period of nine years after the DVD release. Traditional windowing deals like these initially affected the volume of content offered by Netflix's online streaming service, but did not prevent the company's investment in the sector. Although Netflix's still offered physical DVD rentals, subscribers embraced online streaming; by October 2008, estimates showed that 10 to 20 percent of the 8.4 million Netflix's subscribers used the online service regularly.[15]

Simultaneously, changes in licensing agreements began to give way to Netflix's advancement in the market. The company negotiated the distribution of CBS Television Network and Disney-ABC Television Group series in September 2008, ensuring the rights to TV programs such as *CSI:* (2000–2015), *NCIS* (2003–), *Desperate Housewives* (2004–2012), and *Lost* (2004–2010). The following month, a licensing agreement with premium cable channel Starz helped reduce the distribution window of Walt Disney Studios and Sony Pictures films from nine years to just six months.[16] Exclusive agreements with production companies such as Relativity Media, Open Roads, and New Image helped prevent the content from going to competitors such as HBO and Showtime. Two years later, in a nearly $2 billion bet, Netflix struck a deal directly with three major Hollywood studios: Paramount Pictures, Lionsgate, and MGM.[17]

The most important position taken by Sarandos, however, was the investment in a television content distribution window a year after the season originally aired, or the season-after model. Unlike the day-after episodes available on iTunes and traditional syndication agreements, which are generally negotiated after five full seasons, the season-after model created a distribution alternative for broadcast and cable. According to Sarandos, [cable channels] can't really syndicate their shows to other cable channels, and most of the content is so serialized that it's difficult to syndicate at all. We secured exclusive rights to *Mad Men* [2007–2015] partially because we outbid everybody else, but mostly because nobody else wanted it. Because we can get more viewing for that show than anyone else, we can pay more for it than anyone else.[18]

The available options of such content, however, are restricted. According to Sarandos, highly serialized offerings like *Mad Men, Breaking Bad* (2008–2013), and *Sons of Anarchy* (2008–2014), which draw new subscribers to Netflix, are not among the main investments of broadcast networks, mainly because they are as expensive as they are serialized, and therefore and perceived to be difficult to monetize. Meanwhile, pay cable channels such as HBO and Showtime have responded by developing their own subscription streaming services and resisting negotiations with Netflix—a company whose investment in original series almost immediately paid dividends. As Sarandos noted,

Ultimately, we want to produce original content, because it's time we have more control over the shows that matter the most to our customers. We've really come to appreciate the value serialized shows provide. So many people watch them and love them. Our data supports the trend, and that's why you see such an explicit investment in television on Netflix. We've been able to grow the audience for serialized content by recognizing their behavior and securing more and more highly serialized, well-produced, one-hour dramas.[19]

After an experimental co-production with the Norwegian broadcasting network NRK1 for the creation of the mobster dramedy *Lilyhammer* (2012–2014), Netflix's first major solo investment in original programming was the ambitious political drama *House of Cards* (2013–). Sarandos took on the risk of betting over $100 million for a two-season order, a bold but strategic plan in order to profit the symbolic capital Netflix needed to enter the field of television series production. In an interview with *The Hollywood Reporter*, Hastings said, "If we were to get into original programming and it didn't work out, I didn't want it to be because we didn't try hard enough or we weren't ambitious enough.... I wanted to know that if it didn't work, it was because it was a bad idea."[20] The consecration instances proved that it was not a bad idea: produced by and starring Kevin Spacey and written by Beau Willimon, the remake of the 1990s BBC political drama received strong reviews and was awarded the first major Primetime Emmy win for an online television series, given to director David Fincher for the first episode. The series also helped increase Netflix's subscriber base by more than two million in the first quarter of 2013.[21]

However, it was with *Arrested Development* that Netflix made its highest profile investment of the 2013 releases—which also included the successful dramedy *Orange Is the New Black* (2013–), the comedy *Derek* (2012–2014), and the horror thriller series *Hemlock Grove* (2013–2015). Fans of *Arrested Development* anticipated the series' reboot since its cancellation in 2006, especially since the original finale indicated the creative team's desire to return to the Bluth family story. Hurwitz explained the project's transition—initially planned as a film—during the series' hiatus:

> When the show was canceled, I knew I wanted to do more.... At the time, Ron [Howard] didn't think it was a movie, and he was right. When you get a show canceled, it's hard to imagine how you're going to get a movie studio to put money into it. It was really a successful, sneaky thing to have Ron say at the end of the show that maybe it was a movie. That was more of an accident, but I really wanted to make it as a movie at that point.... [Howard] wasn't really on board, but then a couple of years later it started snowballing, and he did start to think it might be a movie. By then, I was doing a number of things, and it was too time-consuming to do it. It wasn't until December of 2011 that I started to really work out the movie. I realized, "Wow, it almost calls for a new form."[22]

At Netflix, the interest in reviving the series was based on the data collected by the taste-based algorithm technology developed by the company to track subscribers' viewing habits. Such technology is a major investment for the company and one of the streaming platform's strongest features, responsible for creating a consumption flow for its subscribers homologous to that of traditional television programming. As Cindy Holland, Netflix Vice President of Original Programming, noted, "[Reviving] *Arrested Development* made sense for us because the show was a cult favorite and we've had for a number of years and knew how many new fans were being created through our service."[23] For Sarandos, Netflix was in a prime position, technologically, to help a series such as *Arrested Development* to find—or rediscover—its audience:

> *Arrested Development* is unique. If all the technology that's in place today were around when *Arrested* came out, it probably would have been a huge hit. Remember, the show was canceled the same year that we started streaming. Prior to that, the notion of catching up on a show didn't really exist. For us to consider [reviving a series], it needs to be more than a great show for the people who love it. We need to try and find a bigger audience for it for the economics to make sense.[24]

In its partnership with Hurwitz and his team, Netflix benefited from the creative power and innovation capacity of the series' producers, which explored the possibility of streaming the full season at once in the design of the fourth season. As Holland noted about Netflix's approach to creative development, "We're buying their vision, not ours. Part of the conversation early on is thinking about it as a 13-hour movie. We don't need recaps. We don't need cliffhangers at the end. You can write differently knowing that in all likelihood the next episode is going to be viewed right away."

The successful partnership for *Arrested Development*'s reboot years after its cancellation arose from such confluence of interests, a harmonic encounter between the plan idealized by Hurwitz to bring back the series—and to expand the innovative characteristic of its language and storytelling—and Netflix's commercial and artistic ambitions to rise within the television industry. The alignment between creative energy and strategic management produces one of the more notable examples of narrative experimentation in contemporary television, evidenced by highly fragmented, complex storytelling and a puzzle-like design of the episodic narrative. However, such innovation and experimentation were already present in *Arrested Development*'s original run, though to a lesser extent—with the interconnection of multiple plots and storylines, the juxtaposition of various layers of jokes (both visually and narratively), and the use of foreshadowing to offer clues about upcoming narrative turns. To understand how the fourth season's storytelling is radically structured, we need to first understand how these elements came to be and were already present in the series during earlier episodes, and how they helped to define *Arrested Development*'s narrative and aesthetic style.

"Now the story of a wealthy family who lost everything": How Arrested Development *Came to Be*

By the late 1990s, Howard's involvement in two very different creative projects inspired his thinking about the possibility of a new sitcom aesthetic. First, he directed *EdTV* (1999), a film about an ordinary guy, video store clerk Ed (Matthew McConaughey), who agrees to have a camera crew following him around for a live, real-time reality TV channel. As the director, Howard led the creative team toward an experimental exercise with the documentary aesthetic and production style of reality television, a key decision that would later influence Howard and producing partner Grazer's approach in defining *Arrested Development*'s own comedic language.

The second important experience at the time was POP.com, a web partnership between Howard's Imagine Entertainment and Steven Spielberg's DreamWorks. The platform offered free short form videos, ranging from one to six minutes, from a variety of genres, but with a particular focus on comedy. With funding from Microsoft co-founder Paul Allen, the project purported to explore web features available at the time, including chat and a content upload option for users; this is a similar structure to what would later become YouTube.[25] Although the company was headlined by notable names in the film and technology industries, POP.com did not survive the dotcom bubble in the early 2000s. Despite this failure, the project strengthened Howard's belief that it was possible to create a more dynamic, spontaneous visual style, or a "lower cost sort of television, that could be produced, that would utilize a new kind of visual vocabulary that was coming out of reality television and docu TV."[26]

In 2002, with *EdTV* and POP.com in mind, Howard and Grazer collaborated with then-president of Imagine Entertainment Nevins to create a sitcom that would explore this new visual grammar—in a quick and affordable production environment. Their planned approach would also eliminate some of the most time-consuming stages of the shooting process—such as filming the performance in the presence of an studio audience—in exchange for a fast-paced, improvised, and self-referential comedic style. According to Nevins, "The intent by Ron, who spent half his life in multiple-camera comedy and half his life as a single-camera director, was to marry the best of both worlds."[27] Nevins played a key role in bringing together the creative talent behind *Arrested Development*. During the 1990s, Nevins worked as a producer at NBC and, among other things, oversaw the development of *Everything's Relative* (1999), a failed sitcom developed by Hurwitz. The sitcom portrayed a family's dysfunctional relationship and, although it was canceled after only five episodes, made enough of an impression on Nevins that he believed Hurwitz to be the voice of a new kind of comedy.[28]

With *Arrested Development*, Hurwitz was very meticulous with the scripts, rewriting the jokes until the last minute. The writing staff regularly visited the set, so that adjustments could be made while shooting with the actors. According to Hurwitz, the goal in writing and refining as many jokes as possible was to ensure the sitcom's realistic style. Such a production process resulted in a densely packed sitcom, with multiple layers of comedy and many set-ups, callbacks, foreshadowing, and complex intertwined narrative threads—characteristics that rewarded re-watching. In addressing such practice as a self-controlled and structured action, Jason Mittell identifies three main aspects: the analytical re-watching, the aesthetic reappraisal, and the social experience. Combined, these aspects produce a more global phenomenon known as "the ludic experience."[29] About the analytical motivation, Mittell says the goal is primarily the close observation of the structure, engineering, poetic, and even the plot presented by the narrative. Creative works such as *Arrested Development*, for example, enable hermeneutic analysis of its episodes, since they encourage an "operational aesthetic of marveling at a show's complex storytelling mechanics alongside the forward drive of the plot."[30]

The content's aesthetic reappraisal is a personal motivation to revisit episodes that trigger specific emotional responses, while the social experience includes the expectation and the analysis of the reactions of newly arrived viewers. Finally, the ludic experience refers to a form of play: "Solving puzzles, seeking patterns, embracing the thrill of discovery, managing our emotional investments, and vicariously experiencing the text through other's eyes. We re-watch as participants in the game, seeking new victories or challenges within the text and our social experiences of media viewing."[31] Mittell also points out that, in comedy series, narrative complexity tends to renew the genres' conventions and to subvert the relationship between multiple plots, creating a tangle of stories that will often intersect or collide.[32] Mittell mentions *Seinfeld* (1989–1998), one of the main references when talking about sitcoms, in which the episodes usually started with four independent threads (one for each of the protagonists) that, throughout the story, would come across each other with unlikely repercussions. According to Mittell, *Arrested Development* enhanced such technique, with coinciding threads intertwined in a manner so complex that the serial narrative became an elaborate set of inside jokes:

> [*Arrested Development*] expands the number of coinciding plots per episode, with often six or more story lines bouncing off one another, resulting in unlikely coincidences, twists, and ironic repercussions, some of which may not become evident until subsequent episodes or seasons. While this mode of comedic narrative is often quite amusing on its own terms, it does suggest a particular set of pleasures for viewers, one that is relatively unavailable in conventional television narrative.[33]

Regarding Howard's original vision of a visual grammar, *Arrested Development*'s style approaches the aesthetics of the *cinéma vérité* observational documentary—something of a "comedy verité" style.[34] Although the series does not explicitly fall within the category of mockumentary television comedies such as *The Office* (2001–2003; 2005–2013), *Modern Family* (2009–), and *Parks and Recreation* (2009–2015), it does employ similar techniques, including the use of handheld camera aesthetics and "archival footage, surveillance cameras, old photographs, and newspaper covers to corroborate or deny the characters' statements."[35] The observational aesthetic in *Arrested Development* also makes it easy to hide visual references and gags in the scenes' composition. According to Christian Pelegrini, such elements are also a cognitive challenge for the viewer, since the dynamic pace of the episodes, with quick cuts and fast dialog, requires a lot of attention. Viewers must keep up with events in the foreground, but technologies like the DVD, DVRs, and Netflix encourage additional consumption to catch self-referential jokes in the background of many scenes.[36] The narrative and aesthetic strategies explored by *Arrested Development* display that the cognitive challenge proposed to the audience was already present in the series' original run at Fox. Such elements have been exacerbated and radicalized in the fourth season at Netflix, as we will observe next.

"Keep those balls in the air!" The Fragmentation of Arrested Development's Narrative Structure at Netflix

As we have argued in this essay, the narrative of *Arrested Development's* fourth season is remarkably different from the way the story was told throughout the first three years of the series, on Fox. Three main aspects were found where it is possible to observe the changes employed in Netflix's episodes, which are explored in detail below. The aspects are: (1) the elevation of the other eight members of the Bluth family to the condition of leading character of their own episodes, as opposed to Michael's role as the protagonist in previous seasons; (2) the circular temporal structure of the narrative discourse; and (3) the shifting balance between individual episodic arcs, focused on each of the nine main characters, and a larger multiperspectivist serial narration throughout the season. Generally, all three aspects identified in the narration of the fourth season relate to an important logistics issue regarding *Arrested Development's* continuation seven years after its cancellation.

With an ensemble of nine actors, all committed to other film and television projects, it quickly became a challenge to bring the full cast back together for an eight-month shooting schedule. According to Hurwitz, this limitation inspired the fourth season's anthological format, with the promotion of all members of the Bluth family to lead character within the context of individual episodes.[37] Although nominal lead Michael remains prominent in the plot—he is the only character to appear in all 15 season four episodes—his attempts to keep the dysfunctional Bluth family together are gone, shifting him into an emotionally codependent father and absentee, resentful brother and son. He has become, in many respects, a man anxious to stay far away from his family. Without Michael's centering presence, the Bluths follow relatively independent paths over the following years (as revealed in their respective episodes throughout season four).

Indeed, the opening credits clearly emphasize the newly shared focus of the series among all lead characters. The original title sequence referred to Michael as "the one son who had no choice but to keep them all together," before introducing the series: "it's *Arrested Development*." Meanwhile, the new credits directly highlight the character who stars at the center of the episode. For instance, Lindsay's (Portia de Rossi) episodes are introduced as "Now the story of a family whose future was abruptly cancelled, and the one daughter who had no choice but to keep herself together. It's Lindsay's *Arrested Development*."

The following chart shows how the main characters are distributed throughout season four episodes (all 2013). Michael, George Sr. (Jeffrey Tambor), Lindsay, Tobias (David Cross), Gob (Will Arnett), and George Michael (Michael Cera) take leading roles in two episodes each along the season, while Lucille (Jessica Walter), Maeby, and Buster (Tony Hale) only star in their own episodes once each.

Episode	Title	Lead Character
1	"Flight of the Phoenix"	Michael
2	"Borderline Personalities"	George Sr.
3	"Indian Takers"	Lindsay
4	"The B. Team"	Michael
5	"A New Start"	Tobias
6	"Double Crossers"	George Sr.
7	"Colony Collapse"	Gob
8	"Red Hairing"	Lindsay
9	"Smashed"	Tobias
10	"Queen B."	Lucille
11	"A New Attitude"	Gob
12	"Señoritis"	Maeby
13	"It Gets Better"	George Michael
14	"Off the Hook"	Buster
15	"Blockheads"	George Michael

In addition to the opening credits, the series' title track plays minor musical signatures with specific instruments for each character. A guitar, for example, plays George Sr.'s musical signature, while Tobias's instruments is a saxophone.

The idea of an anthology series was Hurwitz and his team's solution to developing interesting stories for all characters, despite the cast's limitations with an extended shooting schedule. Season four episodes adopt a character's individual perspective for the duration of six of the seven years that *Arrested Development* was cancelled, in a game of temporal back-and-forth in which the complexity is enhanced by Netflix's content distribution model. Talking about the challenge of putting together such an elaborate, fragmented storytelling format, Hurwitz notes,

> I would say that in its purest form, a new medium requires a new format. You can't do in a short story what you could do for a novel, in a novel. You can't do in a haiku what you would do in a long-form poem. In a perfect world, we would be making something that could be only on Netflix, just like in years prior, you could make something that could only be on HBO.[38]

Hurwitz claims that the company's incentive for creativity was essential in finding the tone and style that shaped season four's episodes and structure. Netflix's prioritizing of the experimental characteristics of a sitcom like *Arrested Development* made Hurwitz's vision for a fragmented and decentralized season possible. As he notes,

Netflix is a very interesting company. These guys are really experimental, fresh thinkers.... I've never had a working relationship like I have with them. I developed a lot of the design of this show with them. That conversation was about, "What are your needs? What are you looking for? Will this work for you guys? Will a show work where you've got one episode per character?" They really were a creative partner. They wanted the next progression of *Arrested Development* and helped me find it, as opposed to telling me how to do it.[39]

Hurwitz also admits Netflix's content distribution system, with the full season's simultaneous release, influenced his storytelling approach to season four: "it's not the same kind of storytelling I think I would have told had the episodes been released through another distributor."[40]

Minding the actors schedule and the season's anthological structure, Netflix's episodes only had all nine actors together for two days of shooting, which correspond to the only two scenes were the cast is completely reassembled in the same place. The scenes are: (1) after the third season finale, when Lucille is arrested, and both the Bluths and their victims aboard the *Queen Mary* are taken to the Harbormaster's Lodge; and (2) at the gathering at the Balboa Towers penthouse, during George Michael's college farewell party before Lucille's trial.

The cast's limited shared screen time is enhanced by the dispersion of scene fragments across multiple episodes throughout the season, punctuated by editing and framing techniques that highlight some characters while obscuring or concealing others. An example with Tobias happens in the scene at the Coast Guard's Harbormaster's Lodge. In "Borderline Personalities," George Sr. and Lucille talk about the matriarch's alibi for sailing off with the ship. In the background, we can see Tobias and Lindsay talking, and Tobias twirls while apparently singing. In the next episode, "Indian Takers," we witness the scene from Lindsay's perspective and find out that Tobias was, in fact, singing: "Oh, is that a gal I see / No, it's just a fallacy," with a pun with the expression "phallus, see." The scene reappears in the Tobias-centric episode, "A New Start," where his singing can be heard in the background of *another* scene—this one featuring a depressed Gob—from "Colony Collapse." Finally, we see the gag once again through Maeby and George Michael's perspective in "Señoritis," as Maeby tries to get attention from her parents by pretending to be dead.

Such an approach, in which necessary information and scene fragments are severed and subsequently placed into multiple episodes, is the foundation of our analysis of the season's circular narrative structure. In the episodes produced for Netflix, the series' diegetic time structure sets aside the primarily chronological way in which the events are portrayed, as it did in the first three seasons, to embrace a centripetal narrative dynamic, a kind of timelessness where the narrative's chronology lapses back and forth at key moments over the episodes. Take for example both the start and end point of the chronological narrative of the fourth season. The story starts immediately after the original series finale, set in 2006, when Lucille tries to escape the police charges against her by fleeting aboard the *Queen Mary*, and follows each of the characters until the "Cinco de Cuatro" celebration in May 2012, covering a period of six of the seven years between the series' cancellation and reboot. However, the events unfolding in both those scenes are dispersed over 13 of the 15 episodes that make up the series' fourth season, and both the starting and the ending point of the chronological narrative are presented in the season premiere, "Flight of the Phoenix."

The Cinco de Cuatro's celebration is the first scene of *Arrested Development*'s season four, displaying an inebriated Michael about to offer sexual favors to Lucille Austero (Liza Minnelli) in order to pay a $700 million loan he cannot afford. However, the story's chronological starting point—Lucille Bluth's arrest on the *Queen Mary*—is only mentioned halfway through the episode, in a flashback covering "The Great Dark Period" in which the series was not aired. The encounters between the main characters at the Coast Guard's Harbormaster's Lodge are portrayed in seven other episodes of the season, always offering new information about the plot, while the Mexican celebration's fateful night, in which the stories of all members of the Bluth family culminate, is shown in eight other episodes. The following chart illustrates the distribution of each of the scenes' fragments throughout the season's 15 episodes. The first column, "Scenes," indicates where the dramatic situation takes place, while the remaining columns refer to the each of the season's episodes. Gray indications represent how many times each of the scenes appeared throughout the season.

Scenes	Episodes														
	01	02	03	04	05	06	07	08	09	10	11	12	13	14	15
Harbormaster's Lodge															
Cinco de Cuatro															

The returning movements to events already portrayed in previous episodes are a specific trait of *Arrested Development*'s fourth season temporal structure and indicate a kind of time experience simultaneity for the series' circular diegetic chronology. Writing of Michael's opening scene in "Flight of the Phoenix," Jaime Nicolás points out:

This first scene will be, for the audience, a beginning, a past, compared with the rest of the character's story, it will be a future for Michael's character—since this is the end of his narrative arc—and, finally, it will be a simultaneous present to the other characters that we will see encountering each other at the same celebration in their respective episodes.[41]

The fourth season's chronology enabled Hurwitz to experiment with a puzzle-like design of the narrative. For example, in the scene at the Balboa Towers penthouse, three months after the *Queen Mary* is capsized, Michael talks to his parents, George Sr. and Lucille, and their lawyer, Barry Zuckerkorn (Henry Winkler), about the upcoming trial of the family's matriarch ("Flight of the Phoenix"). What initially appears to be a conversation among Michael, George Sr. and Lucille (with a quick appearance from Buster) is later revealed to also include Gob ("Borderline Personalities"), Lindsay and Tobias ("Indian Takers"), Gob's fiancée, Ann Veal (Mae Whitman) ("Colony Collapse"), Maeby ("Señoritis"), and, finally, George Michael ("It Gets Better")—as it turns out, the gathering is actually George Michael's send-off party for college.

This example illustrates another way in which season four of *Arrested Development* differentiates itself from the first three seasons: the shifting balance between individual episodic arcs, focused on each of the nine main characters, and a larger multiperspectivist serial narration. Such construction gradually explores the viewpoints and influence of each character to the overall narrative, adding new layers of meaning (and jokes) with every new episode, while clarifying the role of each character to the development of the story. Critic Matt Zoller Seitz emphasizes the importance of Netflix's distribution model while talking about *Arrested Development*'s fourth season.

> Season four plays like a collection of parallel yet interwoven short stories that, when watched in succession, keep revealing new bits of comic business.... When critics write that the streaming model of scripted TV offers new creative opportunities for writers, it's this kind of storytelling that they're talking about: a comic epic made of intricately crafted mosaic tiles that reveal a big picture as you binge-watch.[42]

The night of events at the Century Plaza Hotel, for example, illustrates episodic developments for each of the characters present, while, at the same time, contributing to the season's serial arc. At the time in question, George Sr. attends the rally of politician Herbert Love (Terry Crews) to ask for support to build a wall between the United States and Mexico—thus ensuring that the land he purchased at the border is still valuable ("Double Crossers"). George Sr.'s older son, Gob, on the other hand, is waiting for the baby-faced pop star Mark Cherry (Daniel Amerman) at the Opie Awards, a ceremony that honors young talents from the film and television industries, when he finds out his nemesis, Tony Wonder (Ben Stiller), is doing a magic act at Schnoodle's launch party ("Colony Collapse"). While trying to sabotage Tony's act, Gob accidentally ends up locking Lindsay's boyfriend, Marky Bark (Chris Diamantopoulos), inside the podium where Herbert was going to make his rally's speech, effectively ruining the couple's peaceful protest. The magician's sister, in turn, takes interest in Love and leaves her boyfriend to become the politician's mistress ("Red Hairing").

Maeby and George Michael are also there, since Maeby is the recipient of the Lifetime Achievement Opie Award for her work as a producer at Tantamount Studios. Disappointed to find out that the Lifetime Achievement Opie is only awarded to people whose careers have died, Maeby announces, in her speech, that she is launching an Internet company called Fakeblock, an alleged privacy software George Michael claims to be developing. Maeby, however, does not want their family ties to become public, which in turn leads George Michael to present himself to actress Rebel Alley (Isla Fisher) as George Maharis ("Señoritis"), whose reputation of a Mark Zuckerberg–type of unpretentious internet entrepreneur quickly gains fame ("It Gets Better").

Amidst the character's individual developments, significant pieces from the season's serial arc are also defined: George Sr. influences Herbert to position himself in favor of building the wall between the United States and Mexico, an important point for the Bluths' financial situation over the fourth season. The politician, on the other hand, gets closer to George Sr.'s daughter, and the couple's affair eventually leads to Lindsay's political rise after a rebellion of sorts on the evening of Cinco de Cuatro. This, in turn, results in a Hillary Clinton–like Lindsay running for U.S. Congress in Love's place; finally, Meaby and George Michael's appearance at the Opies is also a key plot point, one that revolves around George Michael's company, Fakeblock, and the George Maharis identity, who ends up becoming Rebel's boyfriend; in fact, the actress is dating another man, George Michael's father, Michael, in what is one of the fourth season's main conflicts.

The fundamental changes in *Arrested Development*'s narrative structure and Netflix's content distribution model encourage specific consumption practices and cognitive demands to fully appreciate the episodes. The act of re-watching is reinforced in order of a more complete understanding of the storylines, since such process requires a high ability to recall and reorganize the previous episodes' events.

Addressing the type of comic construction explored in *Arrested Development*'s fourth season episodes, Jaime Costa Nicolás talks about an "ergodic gag," making reference to Espen Aarseth's electronic literature studies, to describe a kind of narrative in which the reader needs to perform a physical effort to build the text's meaning—ergodic comes from the Greek words *ergon* (work) and *hodos* (path).[43] Nicolás uses the term ergodic gag to reference

situations / planes that separately mean nothing but together they create a comic situation and, therefore, they require a conscious editing process different from that offered by fiction. These grafts which originate comedy are gone and it will be our task to find them in another scene from these 15 episodes. The viewer, therefore, must go through a learning process of all the situations that make up the season to be able to connect both dots. The punch happens between the screen's image and the viewer's imagination.[44]

Once again, there is a highlight to the viewing possibilities offered by Netflix's distribution model: at the streaming platform, subscribers can watch all episodes back to back, pause, rewind, leave an episode and look for a piece of information in another segment, revisit past episodes or scenes, explore future episodes' excerpts, and so on, in a temporality suspension and simultaneity exercise similar to the narrative structure developed by Hurwitz and the series' writing staff.

Conclusion

In "The B Team," in a kindness gesture to help his son Michael, lovestruck by an actress he just met and knows nothing about, George Sr. offers to sign the release form for the movie based on the Bluth family. What appears to be a surprisingly loving moment between father and son turns out to be an intricate favor exchange between the two characters, slowly revealed each time we revisit the scene on the following episodes—"Double Crossers," "Red Hairing," and "Queen B."

What *Arrested Development*'s narrator refers to as "the four-favor family pact" is actually an elaborate gag sustained by the circular narrative structure of the series' fourth season, which reframes the meaning of diegetic past events to contextualize situations shown in the narrative chronological future—a kind of continuous present permanently built in the characters' simultaneous trajectory. Such narrative structure is possible because the agents involved in series development creatively explored the possibilities offered by both Netflix's strategic management and its technological platform. Netflix's executives were aware of its need to associate with the artistic ambitions of a showrunner in order to position itself in a new competitive scenario amongst American production and distribution companies.

In this essay, Bourdieu's theoretical repertoire allowed us to comprehend the relationship between the specific narrative structure of *Arrested Development*'s fourth season episodes and Netflix's interest in changing its position within the television industry. This perspective sheds light on the social agents' decisions importance in both scenarios, because theirs is the ability to comprehend and evaluate the specific historical dynamics in the field of television series production.

It is also important to highlight the skills, wits, and insights of social agents such as Ron Howard, Brian Grazer, David Nevins, Ted Sarandos, and Mitch Hurwitz, in their ability to understand the opportunities and conditions, both creative and marketwise, available at the time in the field of television production. These agents' actions and strategies, when striving for dominant positions within the field, are important to analyze in order to comprehend the logical principles that rule this particular social space of dispute and partnership, competitiveness and collaboration, where the power to define and legitimize what constitutes quality and distinction is constantly negotiated amongst the social agents and institutions. The decisions undertaken by these experienced creators and producers, associated with Netflix's bold business perspective in this market, are essential in understanding not only *Arrested Development*'s poetic and aesthetic innovations, but also the promising current moment within the competitive dynamics of the field of television series production.

NOTES

1. Producer Brian Grazer has confirmed a fifth season for *Arrested Development*, and Netflix's Chief Content Officer, Ted Sarandos, stated that the negotiations with the cast and crew are underway. Elizabeth Wagmeister, "'Arrested Development' Season 5? Ted Sarandos Teases New Netflix Episodes," *Variety*, July 28, 2015, accessed January 30, 2017, http://variety.com/2015/tv/news/arrested-development-new-season-5-netflix-1201549902/.

2. Brett Mills, *The Sitcom* (Edinburgh: Edinburgh University Press, 2009); Jason Mittell, *Complex TV: The Poetics of Contemporary Television Storytelling* (New York: New York University Press, 2015).

3. Examples of *Arrested Development* fan communities that explore the participants' collective intelligence to discuss and discover new information about the series are the Wikipedia page for *Arrested Development*, in http://arresteddevelopment.wikia.com/wiki/Main_Page, the interactive websites that map out recurring jokes and gags from the series, such as *Previously, on Arrested Development* (http://apps.npr.org/arrested-development/) and *Recurring Developments* (http://recurringdevelopments.com/), as well as the series' forum on Reddit (https://www.reddit.com/r/arresteddevelopment).

4. Emil Nielsen, "*Arrested Development*—Behind the Scenes Season 1," *YouTube* video, October 3, 2012, accessed January 30, 2017, https://www.youtube.com/watch?v=PWSaG_hyj9E.

5. Pierre Bourdieu, *As Regras da Arte: Gênese e Estrutura do Campo Literário* (São Paulo: Companhia das Letras, 2005), 245.

6. Jean-Pierre Esquenazi, *As Séries Televisivas* (Lisbon: Edições texto&grafia, 2011); Mittell, *Complex TV*.

7. Esquenazi, *As Séries Televisivas*; Tara Burnett, *Showrunners: The Art of Running a TV Show* (London: Titan Books, 2014).

8. Alan Sepinwall, *The Revolution Was Televised: The Cops, Crooks, Slingers, and Slayers Who Changed TV Drama Forever* (New York: Touchstone, 2013); Gary Edgerton and Jeffrey Jones, ed., *The Essential HBO Reader* (Lexington: University Press of Kentucky, 2008).

9. Michael Curtin and Jane Shattuc, *The American Television Industry* (London: British Film Institute, 2009).

10. David Hesmondhalgh, "Bourdieu, the Media and Cultural Production," *Media, Culture & Society* 28.2 (2006): 211–231.

11. Nancy Hass, "And the Award for the Next HBO Goes to," *GQ*, January 29, 2013, accessed January 30, 2017, http://www.gq.com/entertainment/movies-and-tv/201302/netflix-founder-reed-hastings-house-of-cards-arrested-development.

12. Whitney Friedlander, "'Transparent,' Amazon Break New Ground with Emmy Wins," *Variety*, September 20, 2015, accessed January 30, 2017, http://variety.com/2015/tv/news/transparent-amazon-break-new-ground-with-emmy-wins-1201598229/; Todd Spangler, "Yahoo Loses $42 Million on 'Community,' 2 Other Original Series," *Variety*, October 20, 2015, accessed January 30, 2017, http://variety.com/2015/digital/news/yahoo-misses-q3-earnings-marissa-mayer-narrower-product-focus-1201622483/.

13. Michael Curtin, Jennifer Holt, and Kevin Sanson, ed., *Distribution Revolution: Con-*

versations About the Digital Future of Film and Television (Oakland: University of California Press, 2014).

14. Curtin, Holt, and Sanson, ed., *Distribution Revolution*.

15. Dawn C. Chmielewski, "More Mainstream Movies for Netflix Online," *Los Angeles Times*, October 1, 2008, accessed January 30, 2017, http://latimesblogs.latimes.com/entertain mentnewsbuzz/2008/10/more-mainstream.html.

16. Curtin, Holt, and Sanson, *Distribution Revolution*.

17. Brian Stelter, "Netflix to Stream Films from Paramount, Lions Gate, MGM," *New York Times*, August 10, 2010, accessed January 30, 2017, http://mediadecoder.blogs.nytimes.com/2010/08/10/netflix-to-stream-films-from-paramount-lionsgate-mgm/?_r=0.

18. Curtin, Holt, and Sanson, *Distribution Revolution*, 134.

19. *Ibid.*, 141.

20. Lacey Rose, "Netflix's Ted Sarandos Reveals His 'Phase 2' for Hollywood," *The Hollywood Reporter*, May 22, 2013, accessed January 30, 2017, http://www.hollywoodreporter.com/news/netflixs-ted-sarandos-reveals-his-526323?page=1.

21. Tiffany Kaiser, "Netflix Says 'House of Cards' Is the Reason for Subscriber Growth," *Daily Tech*, April 23, 2013, accessed January 30, 2017, http://www.dailytech.com/Netflix+Says+House+of+Cards+is+the+Reason+for+Subscriber+Growth/article30404.htm.

22. Andy Greene, "'Arrested Development' Creator Mitch Hurwitz on His Two-Year Odyssey to Revive the Show," *Rolling Stone*, May 20, 2013, accessed January 30, 2017, http://www.rollingstone.com/movies/news/arrested-development-creator-mitch-hurwitz-on-his-two-year-odyssey-to-revive-the-show-20130520#ixzz3LiCiMpCI.

23. Rose, "Netflix's Ted Sarandos Reveals His 'Phase 2' for Hollywood."

24. *Ibid.*

25. Emily Farache, "Steven Spielberg, Ron Howard Go POP," *E! Online*, October 26, 1999, accessed January 30, 2017, http://www.eonline.com/news/38906/steven-spielberg-ron-howard-go-pop.

26. Emmy TV Legends, "Ron Howard Interview Part 6 of 6—EmmyTVLegends.Org," *YouTube* video, August 31, 2009, accessed January 30, 2017, https://www.youtube.com/watch?v=q9CYOOpvwKY.

27. Ethan Thompson, "Comedy Verité? The Observational Documentary Meets the Televisual Sitcom," *The Velvet Light Trap* 60 (2007): 70.

28. *The Arrested Development Documentary Project,* directed by Jeff Smith (2013; Los Angeles: 20th Century Fox, 2013), DVD.

29. Jason Mittell, "Notes on Rewatching," *Just TV*, January 27 2011, accessed January 30, 2017, http://justtv.wordpress.com/2011/01/27/notes-on-rewatching/.

30. *Ibid.*

31. *Ibid.*

32. Mittell, "Narrative Complexity in Contemporary American Television," *The Velvet Light Trap* 58 (2006).

33. *Ibid.*, 34–35.

34. Mills, "Comedy Verité: Contemporary Sitcom Form," *Screen* 45.1 (2004): 63–78; Thompson, "Comedy Verité?"

35. Marcel Vieira Barreto Silva, "Sob o Riso do Real," *Ciberlegenda* 1 (2012): 30.

36. Christian Hugo Pelegrini, "Sujeito Engraçado: A Produção da Comicidade pela Instância de Enunciação em *Arrested Development,*" Dissertation, University of São Paulo (ECA/USP), 2014.

37. Greene, "'Arrested Development' Creator Mitch Hurwitz on His Two-Year Odyssey to Revive the Show."

38. Willa Paskin, "*Arrested Development* Creator on the Future of TV and Bringing Back the Bluths," *Wired*, May 2013, accessed January 30, 2017, http://www.wired.com/2013/05/arrested-development-creator-mitch-hurwitz/.

39. Christina Radish, "Mitch Hurwitz Talks Arrested Development Season 4, Bringing Michael Cera into the Writer's Room, and Status of the Movie," January 10, 2013, accessed January 30, 2017, http://collider.com/arrested-development-movie-season-4-mitch-hurwitz/.

40. Michael Groves, "'Chalk One Up for the Internet: It Has Killed *Arrested Develop-*

ment': The Series' Revival, Binge Watching, and Fan/Critic Antagonism," in *A State of Arrested Development: Critical Essays on the Innovative Television Comedy*, ed. Kristin M. Barton (Jefferson, NC: McFarland, 2015), 224–236.

41. Jaime Costa Nicolás, "La Serialidad Ergódica en *Arrested Development*: El espectador/usuario en el Medio Digital," Pompeu Fabra University, 2014, 60.

42. Matt Zoller Seitz, "Matt Zoller Seitz's 10 Best TV Shows of 2013," *Vulture*, December 9, 2013, accessed January 30, 2017, http://www.vulture.com/2013/12/matt-zoller-seitzs-10-best-tv-shows-of-2013.html.

43. Espen J. Aarseth, *Cybertext: Perspectives on Ergodic Literature* (Baltimore: John Hopkins University Press, 1997).

44. Nicolás, "La Serialidad Ergódica en *Arrested Development*," 56.

Circulating *The Square*
Digital Distribution as (Potential) Activism

JAMES N. GILMORE

> Washington may be in an advanced state of moral decay,
> thanks to us Americans mostly disengaging from civic life.
> But ordinary Egyptians are in a state of total & complete
> civic engagement. what [*sic*] they are doing should be an
> inspiration to all of us here, who are increasingly feeling
> defeated by the nexus of money, power, and corruption that
> is slowly corroding the core of America.
> —Anonymous review of *The Square* (2014)

The early stages of many protest movements of 2011—including the
Egyptian Revolution, the Arab Spring, and Occupy Wall Street—were predicated on digital media access and the ability to disseminate and circulate
various calls to action. As Tim Markham notes, "Facebook was particularly
popular in the Tunisian uprising, Twitter was the medium of choice in an
already well-established culture of blogging in Egypt, while online the civil
war in Syria is largely being played out on YouTube."[1] Social media was integral in creating communities and organizing protests. These users set dates
and places for physical communal engagement as well as spread the rhetoric
of their social action across global virtual space. These social movements
immediately—and increasingly—became the subject of popular histories and
academic studies that, to varying degrees, detail how digital media fosters
political revolution as well as how civic engagement operates in conjunction
with the ideals of participatory culture. Much as how Twitter helped organize
protests, it also played a role "as a bridge between Egypt and outside communities."[2]

This essay attends to the mediated afterlife of these revolutions in documentary images, arguing that the digital circulation of political documen-

taries in on-demand media spaces such as Netflix is crucial for continuing to emphasize social media's potential to incite social action. Precisely, I am interested in the documentary *The Square*, directed by Jehane Noujaim, which was partly crowd-funded, premiered at the 2013 Sundance Film Festival, and was subsequently acquired and distributed exclusively through streaming media platform Netflix in early 2014.[3] As the first political documentary acquired exclusively by Netflix—which heretofore had largely dealt in original series programming and acquisition of older, previously released media— for a streaming release, the documentary's distribution site and method of circulation became as much a story as its images of political revolt. The film's 2014 Academy Award nomination for Best Documentary Feature lent it a sense of cultural capital that, coupled with its discursive position as a major step forward in Netflix's plans to acquire and disseminate exclusive content, frames it as an important component of current political documentary distribution. Apart from how it both represents political revolution and reflects the corporations that digitally shepherd documentaries to subscribers, *The Square* must also be understood in terms of how it is both about and reflective of broader dynamics of media spreadability.

Political documentaries made about these movements and distributed through these sites espouse democratic ideals in that they circumvent the structures of theatrical distribution to ideally reach wider audiences, but these implications are far more complicated when analyzed through problems of media circulation, such as geoblocking. As Netflix begins to acquire an increasing number of social and political documentaries, its site will simultaneously become a mobile archive of cultural, historical, and social representations. Along similar lines, W.J.T. Mitchell has argued archives are not only physical institutions; they are increasingly online discursive channels where "the shaping of perceptions of history ... is immediately represented in audio-visual-textual images transmitted globally."[4] Considering that Netflix now accounts "for 34.2% of all downstream usage during primetime hours" in North America, we must speculate not only on what its rampant consumption means for reshaping film and television into what Chuck Tryon calls "on-demand culture," but also what is to be gained from all of this consumption, especially as Netflix positions part of its output as a boutique documentary distributor.[5] Our research questions must focus on who accesses the site, why viewers engage particular texts, what they derive from consumption, and the sociopolitical implications of on-demand culture.[6] More than promoting certain forms of consumption, analyses of Netflix must consider the political potential of this platform for raising awareness and education about many key global social issues.

This essay considers how this mode of media distribution and circulation both represents and imagines certain forms of civic engagement. *The*

Square's history of the Tahrir Square protests capitalizes on multiple hallmarks of participatory culture's shift towards do-it-yourself (DIY) or democratized media production, as outlined most notably by Henry Jenkins, who suggests participatory culture has engendered less "changes in institutions or laws, which are the focus of traditional political science, but more ... changes in communications systems and cultural norms," most notably in the practices of popular culture.[7] I argue that the dissemination of political documentary content through Netflix highlights the political potential of media spreadability, and the need to continually regard the politicization of media circulation and social media. The act of viewing and discussing *The Square* is, in this analysis, part of ongoing global conversations occurring across digital technologies and through grassroots media efforts to contest social and political inequalities. Media objects like *The Square* are partly mimetic calls to action and partly consciousness-raising projects that link locally situated social movements into a globally fluid network of media dissemination, consumption, and activism.

This essay proceeds first by discussing some of the spatial stakes of social media use in relationship to activism in general and the 2011 social movements in particular as well as the evolving place of political thought in literature on participatory culture, before turning back to *The Square* to analyze its aesthetic and its reception. Ultimately, I argue that studies of participatory and on-demand culture can learn significantly from each other in terms of how media circulation services such as Netflix can further the political potential of documentaries through their ubiquitous mobility.

Bridging Online/Offline Space

The social movements of 2011 have already inspired a number of studies that have emphasized, in varying degrees, the role of social media use in facilitating and organization physical protests against the state and its policies.[8] Importantly, the *fact* of social media is not enough in discussing these movements. Mark Warschauer argues, "What is most important about ICT (information and communication technology) is not so much the availability of the computing device or the Internet line, but rather people's ability to *make use* of that device and line to engage in *meaningful social practices*."[9] Much like the mere existence of Twitter in Egypt is not enough to create social action, neither is the existence of *The Square* on Netflix evidence of its importance for continuing projects of social engagement or for raising consciousness about the events in Egypt. Rather, attention must be paid to the text, its circumstances of distribution, and the way users imagine use value for their media consumption. Following Joel Penney and Caroline Dadas's analysis of Twitter use in Occupy Wall Street, which aims to discover "how people are using language to construct new social and political realities, and how they are incorporating social media technologies into that process," this essay considers how Netflix fits into the imagination of social and political realities, and how its standard modes of consumption could foster social action.[10]

Jan van Dijk similarly takes aim at those who suggest that access to technology alone is able to operate as a corrective to social problems. Van Dijk importantly thinks about media access in terms of spatial relations, arguing, "not having access to online environments increasingly also means absolute exclusion from particular offline environments and from a number of social, economic, and cultural opportunities."[11] In this kind of spatial access, communities formed online can translate into socially and politically beneficial "offline environments"—what might be understood as a digital grassroots activism working to bridge different kinds of spaces. As Manuel Castells argues, "There is no question that the original spaces of resistance were formed on the Internet" as a way to mitigate against the difficulties of "traditional forms of protest."[12] Digital spaces afford a place for protestors to find groups and to discuss their aims and tactics before proceeding into physical space. These models echo Henri Lefebvre's famous contention that "(social) space is a (social) product," that part of how we collectively apprehend space is through the social relationships acting upon any given site—so too did the social connections of Twitter and Facebook mesh with and transform the sociopolitical potential of Tahrir Square.[13]

In Egypt, those who responded to the digital calls for participation were overwhelmingly young, "for whom social networks and mobile phones were a central part of their way of life," prompting recognizable age disparities between those who can not only access but actively use social networks efficiently.[14] Paolo Gerbaudo reads the spatial differences between social media and protest camps as rethinking how we imagine the space of activism and revolution.[15] Here, social media participate in a "choreography of assembly" designed "to sustain their coming together in public space."[16] Again, social networks bridge divides between "virtual" (or online) and "real" (or public) space. They form a directive that creates methods of behavior and modes of being that can translate into densely packed protest camps that rewrite public space in the name of ideological revolution.

That is to say, the 2011 social movements—and their subsequent representations—create a new form of "MediaSpace," Nick Couldry and Anna McCarthy's dialectical term "encompassing both the kinds of spaces created by media, and the effects that existing spatial arrangements have on media forms as they materialize in everyday life."[17] Utilizing this concept suggests that, as much as social media movements have instigated political protests in physical spaces, so too has the everyday circulation and consumption of media representations through Netflix added another level to the kinds of spatial imaginaries we might construct. Manuel Castells further adds to this discussion of what "space" means in this context. In discussing public spaces that have been co-opted by revolutionary bodies, he asserts, "Occupied spaces are not meaningless: they are usually charged with the symbolic power of invading sites of state power."[18] Where van Dijk partly suggests analyzing digital media access through how online participation translates into or inhibits participation in offline spaces, Castells complementarily notes that the very ability to occupy a space, and to do so through online social organization, turns Internet users into producers of media messages capable of inventing new ways of everyday being.[19] This conjures images of media spreadability— or media's ability to circulate in a number of different ways in digital space. The ability of media to flow across space means that public and online spaces become increasingly interrelated, such that activism "extends from the space of places to the space of flows."[20]

This space of flows is important, and as Philip N. Howard and Malcolm R. Parks note, "there is a connection between technology diffusion, the use of digital media, and political change. But it is complex and contingent."[21] Of course, we should be wary of overstating the importance of digital communication technologies. As Zeynep Tufekci and Christopher Wilson have argued in their study of the Egyptian protests, "Nearly half of those in our sample reported that they had first heard about the Tahrir Square demonstrations through face-to-face communication," while Facebook accounted for 28 percent of meaningful communications, and only 1 percent cited texting, e-mail, or Twitter. Interpersonal communication, for Tufekci and Wilson, continues to be a dominant and affective means of communication.[22] Merlyna Lim pushes back on the label of "Facebook revolution" to stress the relationship between online activism and those who work to promote it, arguing "the power of networked individuals and groups who toppled [the] Mubarak presidency cannot be separated from the power of social media that facilitated the formation and the expansion of the networks themselves."[23]

One of the most-cited instances that sparked the Egyptian revolution was Asmaa Mahfouz's January 18 Facebook vlog, which was later uploaded to YouTube and subsequently spread through social networks and other forms of virtual sharing. Mahfouz's video makes clear that the Internet can *spread* social revolution in profoundly new ways. As if mirroring the event it represents, so too does *The Square* spread digitally—not only is it viewed through Netflix or through illegal torrent downloads, but information about it is also shared, liked, retweeted, and posted across a number of social media platforms. Much as the documentary itself attempts to raise consciousness about the events in Egypt, so too do the tools of social media allow viewers and users to promote the documentary through social media networks.

Henry Jenkins's ongoing work on online participatory culture is instructive for demarcating and tracing the implications of media's spreadability. As opposed to media distribution, which may send a media text through one channel and in one form, the idea of spreadability allows an object to be changed and appropriated for a number of contexts.[24] This complicates how we conceive of viewing *The Square*, as it is ostensibly only "available" on Netflix yet also downloadable through torrents; it is both distributed and spread. As Jenkins, Sam Ford, and Joshua Green assert, "This shift from distribution to circulation signals a movement toward a more participatory model of culture, one which sees the public ... as people who are shaping, sharing, reframing, and remixing media content" in new ways.[25] While *The Square* has not been "remixed"—users have not yet reedited its digital images—it exists, like many political documentaries, to be spread.

Spreadability is, then, "the potential—both technical and cultural—for audiences to share content for their purposes," where sharing is amplified by social media platforms that emphasize items like "the embed codes that YouTube provides, which make it easier to spread videos across the Internet, and encouraging access points to that content in a variety of places."[26] The aim of spreadability, if it circulates through enough channels, is to move "audiences from peripheral awareness to active engagement."[27] Beyond acts of re-circulating, spreadable media allows users to become participatory and active in ways that range from sharing and clicking hyperlinks to moving into actual physical spaces.[28] The move from awareness to engagement is thus analogous to the movement between online and offline spaces. Joseph Turow has nuanced the importance of the hyperlink by calling attention to its connections to corporate tracking and other modes of monitoring that entail the "industrialization" of the link by quantifying clicks and shares for things like advertising.[29]

The notion of spreadability relies on the human capacity to spread media through hyperlinks, clicks, shares, and other emergent forms of digital communications tools that connect the Internet. This stands as a sharp contrast to another oft-deployed metaphor of digital culture, virality, which for Jussi Parikka says, "expresses such key tendencies of network culture as communication, self-reproduction, transmission, and de- and reterritorializing movement."[30] Drawing on Steven Shaviro, Parikka contends, "Selfhood is increasingly depicted as an information pattern, where the 'individual' becomes merely a host of parasitic invasion by information capitalist patterns of repetition."[31] If media spreadability envisions a potential for subjects to create communities through communicating meaningful media to each other, this model of virality sees digital culture as wholly disempowering, where the self gradually disappears. That is to say, a debate about the terms used to describe how media and selfhood work in digital culture is also a debate about what the self can do in a politically empowering way. As much as media circulation can spread in newly democratic ways, so too is it controlled at corporate and policy levels in ways that significantly effect how we might use it socially and, importantly, democratically.

Netflix's content cannot be freely circulated. Although users can give passwords to each other or implore other Netflix users to seek out particular texts—not to mention circumventing Netflix entirely through illegal downloading—it still strives to control the flow of spread and circulation. When the content is political, both its movement through space and the barriers to its spreadability take on political dimensions. User decisions to engage *The Square*, similarly, become political, such that "audiences play an active role in 'spreading' content rather than serving as passive carriers of viral media: their ... actions determine what gets valued."[32] Spreadability, then, is a crucial step of participatory culture, where "the erosion of traditional boundaries," such as those between "mere" fans and "political" activists, becomes central.[33] Even as this boundary is unsettled by digital distribution and the ability to circulate media images, texts, and ideas, Netflix and *The Square* are still somewhat positioned in a traditional model of media delivery—on-demand, certainly, but nevertheless a form of controlled consumption, to borrow Lefebvre's phrase.[34]

Digital inequalities have social consequences for spreadability's actual capacity to enact grassroots political activism—to enact movement across different kinds of space. As Jenkins, Ford, and Green note, spreadability "has made it easier for grassroots communities to circulate content than ever before, yet the requirements of skills and literacies, not to mention access to technologies, are not evenly distributed across the population."[35] This call for media literacy harkens back to Warschauer's analysis, and has also been at the heart of the Egyptian revolution and other global protests' efforts to generate international attention and conversation about their work. If "acts of circulation shape both the cultural and political landscape in significant ways," then *The Square* generates even more complex problems; its circulation is not tied to an explicit call for political action, but to a documentary whose ostensible goals are to teach, inform, and enlighten. American viewers, to use one hypothetical example, can draw parallels between Egypt's situation and Occupy Wall Street—or any other extant or nascent activist movements—potentially using the film as a template for their own forms of activism. It is not enough to merely discuss the mobility and fluidity of media; we must strive to understand *where* it circulates and *to whom*.

At the heart of this discussion is a question of how we might consider *being at a computer* a form of activism, and whether or not the consumption and spread of media texts is a kind of user engaged, politically motivated, and ultimately effective way of raising consciousness, shaping discourse, and—hopefully—altering social or political landscapes. To put it another way, "an individual who 'productively' responds to one media property, brand, or cause may be a 'passive' listener to many others; activity and passivity are not permanent descriptions of any individual."[36] Certainly, Netflix's efforts at controlled consumption attempt to structure certain modes of activity of access. Ganaele Langlois has similarly argued that despite the democratic potential of online participatory practices, we must also attend to notable cultural paradoxes; namely, we must "identify processes of governance" that articulate humans, technologies, and processes to each other to render the sense of stability in online social environments.[37]

Michel de Certeau's distinction between uses and tactics, where the former is imposed by some kind of power structure and the latter represents how users actually act within that structure, is useful here.[38] The models of governance Langlois identifies fit into the ideological structures that in some way determine how media spread and are used. In contrast to spreadable media, Rita Raley invokes the term "tactical media" to define "the aesthetic and critical practices that have specifically emerged out of, and in direct response to, both the post-industrial society and neoliberal globalization."[39] Raley looks at more politicized "digital art practices" to consider how new media technologies have created a space for digital artists to mobilize their politics. Tactical media then "signifies the intervention and disruption of a dominant semiotic regime," drawing on de Certeau's sense of how users navigate an imposed structure.[40] Tactical media operates briefly—its use is not protracted—and there are strongly ambivalent questions about whether this kind of political work can affect broad social transformations; de Certeau argues, "There are no proffered fantasies of radical systemic change: it exists as a possibility within the realm of the imagination—another technology of simulation—but it requires collective action."[41] As de Certeau further claims, tactics give up what they win; they offer "mobility, to be sure, but a mobility that must accept the chance offerings of the moment."[42] Raley's analysis speaks to how art and media operate as critiques that might create sparks leading to social transformations. Different from the consciousness-raising projects of shareable and spreadable media, where education and ideas are often prized, these are more radical and politically subversive works that aesthetically, politically, and ideologically challenge the hegemonic constructs of neoliberalism. They offer "a more fluid, extensive, and thereby more powerful set of art-activist practices."[43]

Raley's work on political performativity may seem counterintuitive in a discussion of *The Square*. She establishes, however, a crucial lens for what it means to make media that spreads. *The Square* was not explicitly produced as a text to be circulated digitally—it initially premiered at Sundance and Netflix bought distribution rights after several festival screenings. In a sense, *The Square*'s relatively conventional aesthetic and structure may mark it as not pushing the politicized bounds of Raley's tactical media model. I want to argue the act of digital distribution *itself* is a political performance. Marking it as such fosters connections between the Egyptians shown watching Facebook and YouTube videos in the diegesis and the viewer watching *The Square* from a laptop or tablet from a number of geo-locations. The potential links between these two subjects—film viewer and diegetic video watcher—point to the hope for this kind of media to continue tactical proliferation. That is to say, the film envisions a world where tactics do not give up what they win, but can have broader transformative impacts.

Beyond the hopeful, tactical conceptions of spreadable media favored by scholars like Jenkins and Raley, Evgeny Morozov's concept of "slacktivism" provides a useful counterpoint to the potential bridges between online and offline space. "Slacktivism" is the "digital sibling" of activism—a way of engaging causes that "makes online activists feel useful and important while having preciously little political impact."[44] Slacktivism is online activism that cannot—for any number of reasons—bridge into offline space; it cannot enact "real" social change. Morozov takes technological determinism to task, noting the actual effects of technology—especially in relation to their ability to empower or liberate the individual user—"were often antithetical to the objectives their inventors were originally pursuing."[45] The quick satisfactions slacktivism offer gets in the way of "risky, deep, and authentic … commitments."[46] The danger, then, is that the bridge between online and offline space remains forever unbuilt.

The promotional material for *The Square* heavily favors the offline spaces of protest, while also favoring Netflix's position as an online streaming service rather than the tensions between online and offline activism in the protests themselves. The official trailer, released on YouTube one month before the film's premiere, does not include any mention of social media, although there are several shots—including the trailer's final shot—taken from amateur video cameras. Much of the posters and images circulated online and elsewhere featured the tagline "The people demand the downfall of the regime," focusing on the revolutionary thematics of the film. The day of the Oscar nominations—one day before the film's release—Netflix took to its social media accounts to usher a new tagline: "Oscar nominated today, streaming tomorrow." This sort of rhetoric reframes the documentary towards the availability of on-demand culture and the promises of media circulation.

The promotion of *The Square* across social media platforms supports Netflix's desired strategy of using social media to encourage product consumption, but taglines like "Oscar nominated today, streaming tomorrow" can also suggest Netflix rather cynically co-opted the political power of circulation for capitalistic gain. This view omits any kind of political empowerment that may come as a direct or residual effect of consuming and sharing this media. One of participatory culture's many goals might be understood to be the creation of new channels that instill a more democratic media culture and, in turn, a more democratic political culture. In a discussion between Nico Carpentier and Henry Jenkins on this democratic potential, Carpentier suggests, "Democracy and participatory culture will always be unrealized.... There will always be struggle, there will always be contestation."[47] Democracy and participatory culture are dual ideals, but for Carpentier, this utopian thrust is part of what gives them their power. He implores us to realize the impossibilities of this culture without straying from its immense potential: "Participation allows for the performance of democracy, which is deemed an important component of the social in itself."[48]

The production of *The Square* also reveals a participatory, democratic ethos. Shot by an Egyptian and with the help of Egyptian subjects, the film serves in part as a model for what this mode of on-the-ground, localized independent filmmaking might accomplish in representing social revolution. David MacDougal's emphasis on creating empowered subject voices, and Thomas Waugh's concept of a "committed documentary" that works with politically motivated communities are two crucial intersections from the field of documentary studies that again emphasize the importance of individual voices in creating politically powerful representations.[49] In *The Square*, Noujaim's filming is as much a part of the revolution as she is an observer of that revolution. She further constructs subjectivities by screening portions of her film to revolutionaries throughout her editing process and incorporating their responses; in her words, "What the film does … is it humanizes the struggle of Egyptians and really shows what the human story was behind those scenes."[50] Her work with her community constructs civic media as representing and aiming to inspire particular forms of political action. The goal of this distribution strategy is to actively circulate "a greater diversity of perspectives" and, in turn, "motivate participation in the political process."[51] As the site of distribution, Netflix subscribers could—through the educative aims of something like *The Square*—work towards better policy decisions for social equality when given civic opportunities. In short, beyond its desire to be spread and shared in order to educate, this documentary has a *tactical* aim.

The Square *in the Rectangle*

As it has acquired social and political documentaries, Netflix increasingly functions as an archive that can spread socially progressive political ideologies. Tryon's work thinks both about digital deliveries and circulations of media writ large as well as Netflix as a specific entity bound up in the wider web of cultural engagement digital circulation engenders. As he argues, digital delivery is both a promise of "ubiquitous access," but also instills a more individualized mode of consumption, where film-going in particular loses its social import.[52] Tryon introduces "platform mobility" and "resistant mobilities" as two disparate ways to understand how texts circulate in digital space. In platform mobility, "it's not just texts that circulate. So do screens," and viewers can seamlessly "move" media across devices, creating an "individualized, fragmented, and empowered media consumer" who makes her own pathways through the media she selects.[53] While Tryon's discussion of platform mobility nuances the relationship between online and offline spaces—especially as online media can "move" through or inhabit increasingly more offline spaces—he importantly considers systemic constraints barring how users can consume texts. For instance, geo-blocking, where texts are available on Netflix in one country but not another, leads to alternative methods of cultivation and consumption, where users must use extralegal measures to acquire content. Tryon defines these as "resistant mobilities," or "activities that defy the practices promoted by the entertainment industry."[54] The problems of geo-blocking are crucial to understanding the circulation of *The Square*; although "it's been released in more than 40 other countries, it did not receive distribution in Egypt until June 2014, nearly half a year after its US Netflix debut."[55] Much as Tryon talks about "resistant mobilities" in individual as opposed to communal terms, larger media entities—in this case, YouTube—have stepped in on behalf of *The Square* and allowed it to circulate digitally. The film received "an exclusive YouTube release in its home country of Egypt, despite not yet being approved by official censors."[56] YouTube, in turn, geo-blocked the film for every country *except* Egypt to encourage others to view it through Netflix. Here, geo-blocking and resistant mobilities work together to create access and circulation that make the media more widely accessible.

The other major wrinkle to these practices is the very emergence of algorithmic culture itself, where Netflix recommends and acquires certain kinds of movies based on user preference and accrued data about viewing and rating habits. As Blake Hallinan and Ted Striphas have argued, Netflix's recommendation system points to "a court of *algorithmic* appeal in which objects, ideas, and practices are heard, cross-examined, and judged independently, in part, of human beings."[57] Their suggestion is an important one, for it signals that the curation and promotion of certain kinds of material to certain kinds of individuals with matching taste profiles. In this model, the promotion of *The Square* becomes less about raising consciousness across demographics than reinforcing certain sociopolitical ideologies held across certain demographics of Netflix consumers.[58]

At this juncture, I turn to the text of *The Square* to think about the way it frames digital media spreadability and circulation as foundational to the Egyptian revolution. This section analyzes how these practices are aestheticized for a more prescriptive and far-reaching account of how media spreadability functions in relation to social revolution. I argue that the film's title is actually a multi-layered signifier. Ostensibly, *The Square* references Tahrir Square, the central space of the Egyptian Revolution. "The Square" also represents multiple media frames—both the *actual* frame through which the user watches the film as well as the plurality of frames that populate the film's opening act, embedding websites and YouTube videos within frames. *The Square*'s opening invokes a nesting-egg aesthetic that reflexively displays the act of watching digital media. This analysis follows Thierry Kuntzel's neoformalist suggestions that the openings of films are key to laying the aesthetic and thematic groundwork for the rest of the text.[59] In locating an aesthetic representation for how digital media access builds bridges between online and offline communities, I turn to the opening sequence of the film leading up to and just past the title card. The first shot of *The Square* is of a power outage across an urban skyline; the subsequent shot is a close-up of a match being struck and a candle being lit. At once, the film sets up its dynamics of space, oscillating between the conflicts of the wider city and resistance being offered at the individual level. This movement from establishing shot to extreme close-up brings the film to the local level, but also substitutes one form of power for another—not just the industrial electricity for the candle, but a governmentally controlled grid for the individuals in the room finding their own light source. As if the metaphor of light across these shots were not strong enough, one of the first lines of dialogue is "The lights are out all over the world. The lights are out all over Egypt." Again, this operates as a symbolic extinguishing and igniting—the "light" is not only the actual electricity, but also the structures of power that control it and the people who are left "in the dark."

From here, *The Square* introduces one of its protagonists, Ahmed, who is photographed walking in the streets and in his apartment. After several moments of watching him discuss his life in voiceover as he moves through these everyday spaces, *The Square* begins a somewhat extended sequence of Ahmed in front of his laptop. It starts with an image of a YouTube video titled "torture in egypt [*sic*]." The camera photographs the computer monitor, somewhat distorting the image, but at the same time providing a more indexical record of what was circulated and how it would look within the space of one's personal computer. The following shot cuts to a close-up of Ahmed's face with the back of the laptop occupying the bottom third of the frame. This establishes a shot-reverse shot pattern between Ahmed's face and the laptop screen, where the physical laptop itself never leaves the frame; it is either what we are looking at to see the videos, or occupies a small part of Ahmed's close-ups. The sequence continues by showing another video—this one titled "Egyptian police torturing a woman murder suspect 1"—with the view counter showing upwards of 511,000 views. From here, the sequence cuts to a Facebook photo of a young man; the mouse clicks to the next photo in an album, revealing the same man's face severely beaten, bloody, and broken. This image—as seen on the edge of the frame—has 70 likes and 11 shares.

When the sequence again cuts to the laptop, Ahmed is now watching the Asmaa Mahfouz video mentioned earlier in this essay. Mahfouz says, in the video, "We will go down and demand our fundamental human rights." The sequence cuts to black as the video concludes, but in the next shot, Ahmed's voiceover remarks, "I went to the streets. I found everyone around me felt just as I did." Images of a moving, ever-expansive crowd capture people taking photos and videos with cell phones, and a moment later the sequence cuts back to Ahmed in his apartment on his own phone, telling someone to "tell everyone to come down to the streets." This footage continues for some time, concluding with cell phone footage surveying the square, now filled with people. A man off-screen yells, "we have taken the square," and the sequence cuts to an overhead shot of Tahrir Square completely filled with people. The title card—*The Square*—superimposes over this overhead shot. The people, notably, form more of a collective circle, such that their geometric figuration from above actually operates as a rebellion against the shape of the space.

This sequence, while only comprising approximately the first five minutes of the 105-minute documentary, aestheticizes the entire logic of a social revolution based on digital media access and circulation. It individualizes the act of media consumption through conveying the YouTube and Facebook videos as shots from Ahmed's perspective, and it nestles them within the frames of the laptop to show how they look on particular forms of technology—a technology on which viewers might also be watching *The Square*. The lighting of the match allegorizes the ability to spread, share, and consume this media; it is no accident that the numbers of views, likes, and shares these media have received populate the corners of the frame. The form of digital media is displayed not only to contextualize how an Egyptian saw these images, but also to recognize how others view media on Facebook, YouTube, and other digital platforms. This opening foregrounds the shareability of digital media and, in showing Ahmed both at home and in the streets, how they operate to transform the relationship between offline and online spaces. The overhead shot of the crowd gathered in the Square, then, is the demonstration of how online spaces rewrite public spaces.

Having discussed *The Square*'s visual aesthetic, the site of its primary distribution, and the potential benefits of spreading this media for raising political consciousness, I turn to a discursive analysis of the film's reception. While the discourse surrounding Netflix's acquisition and distribution of the film focuses more on raising the website's cultural capital through its ability to circulate premiere global independent filmmaking, Netflix users' reception of the film suggests an array of everyday uses for largely American-identifying viewers. In creating parameters for a reception study of *The Square*, I surveyed the user reviews that had been left on the film's official Netflix page. At the time of this writing, approximately five months after the film's premiere, it has generated 133 user reviews. Netflix reviewers are not required to provide any identifying information, so I am unable to assign names, genders, or geographical locations of any certainty to these reviews. The selected reviews attempt to retrieve a sense of how this film is being received *politically*—what do users see as valuable to its consciousness-raising project? What political or social effect might it provoke? From the 133 reviews, I focused on those where the reviewer self-disclosed information about their nationality—overwhelmingly, they were either American or Egyptian, which may admittedly have much to do with my subscription to the American version of Netflix's browser. These reviews paint a picture of how viewers affectively respond to the film, understand it as an education endeavor, and see it as a piece of political provocation. Although the bridge between online and offline spaces of activism is not often explicitly discussed in these reviews, they nevertheless point to the affective dimensions of the film's response and the ways users may continue to use it in their political—and perhaps public—life.

For instance, numerous reviews tie *The Square*—quite remarkably—to conservative U.S. gun policy. Most potently, one reviewer suggests: "This [documentary] also reminds you for those living in America why the 'right to bear arms' is there for that sole purpose, to allow its citizens to fight back with maximum effort against a government that tries to do this to their citizens. Every person who thinks people should be disarmed should watch this movie and learn a powerful lesson."[60] Another saw the film as arguing it "shows exactly why people must be armed to be free. Hopefully when the time comes in america [*sic*] we will band together, take up arms, and take back our country," and still others saying, "I fear this very thing may happen in the US."[61] And again: "this is what happens when disarmed citizens cannot defend themselves from their government."[62] In a sense, this is an odd projection of one country's political climate onto another, yet it also speaks to the polysemous ways audience members negotiate the meaning and politics of cross-cultural media.

The majority of reviewers who identify with an American national identity are not trying to make these kinds of ideological connections. They rather see *The Square* as a way to redefine their own sense of what "democracy" and civic action mean. Like the quote that serves as an epigraph to this essay, *The*

Square reminds American viewers of the importance of civic engagement. As another reviewer commented, "Americans in particular need to see this. We love to lecture the rest of the world about how democracy is supposed to work. Well, this film shows real genuine democracy in action. And surprise, it is messy and chaotic."[63] Another echoes: "As a young American, documentaries like this are needed so people across the globe can understand what is happening outside our personal 'bubbles' in which we live in."[64] The documentary serves two purposes here: to remind American viewers that U.S. democracy is not indicative of democracy the world over, and to remind Americans, whose elections continue to have notoriously low voter turnouts, of the importance of civic engagement.

Many reviews center on *The Square*'s educative potential. Reviews regularly state, "I feel like a learned a lot" about Egypt, or "I realized how ignorant I was as an uninformed American."[65] Still others see its boots-on-the-ground perspective as inherently more historically valuable: "Every American who has no idea about the *truth* of Egypt, who judge by what our media wants us to know as truth, should watch this wonderful, amazing documentary," or "It gave me a full understanding of what happened at Takir [*sic*] Square."[66] One viewer who self-identifies as an Egyptian who "lived" the revolution argues, "I think this is the closest we can get to truth before historians alter it in favor of whatever regime happened to be at the moment of them writing it."[67] Another who self-identifies as American suggests, "This should be shown in every modern history or political class in every college in every country."[68] The importance of the film comes in how it stands as an appropriate model for telling a history of revolution, taking the perspective of those on the ground instead of those in power. The "closeness" of this documentary to the event itself, instead of clouding its ability to dissect the revolution, is rather the real benefit—it is a document *of* the revolution as much as it is a documentary *about* the revolution.

Of all these responses, the most interesting and most useful for this analysis are those that incorporate some kind of revolutionary rhetoric. For example, one states, "This film motivates me to be the change I want to see. Go out there and change the world. Don't wait for the world to change you."[69] Others suggest *The Square* is "a handbook for change,"[70] providing methods that could be applied to a number of social circumstances for enacting revolution. Still others discuss, in impassioned terms, how the film affected their personal worldview: "I never delve into foreign politics, but NOW I know that I MUST!! This Doc, shows how YOU can actually manifest CHANGE in your Governments by the choice of A TRULY FREE PEOPLE!"[71] Another reviewer even went so far as to say it "made me want to travel to Egypt and join in the movement for peace, justice, and democracy," and that the film provides lessons that "are extremely valuable to revolutionaries globally."[72]

Whether they focus on their individual response to the film or a belief that it can affect groups and civilians the world over, these reviews focus on the *fluidity* of its circulation and its ability to digitally "travel" across contexts. While stopping short of calling *The Square* a universal text, these reviews all stress the film can motivate and enable social revolutions "globally"—that it has considerable value apart from educating viewers about the Egyptian Revolution.

I have taken the space to survey these reviews—despite the inability to truly position their authors geographically or politically—because they reveal much about the polysemous ways users approach, view, and discuss *The Square*. The viewer who suggests moving to Egypt to join the revolutionaries, for instance, may not actually do so and may forget about the film after writing her brief review, but the suggestion that the film works across historiographic, educative, political, and activist axes is explicit in this sample of reviews. Further, the global circulation of *The Square* suggests the meanings of political documentaries are radically shaped by where and when they are viewed, such that the increased "mobility" and "spreadability" of media becomes important in relationship to the individual or group consumer. This is to say that American viewers ostensibly project their own politics and circumstances into the film, attempting to create analogies—appropriate or not—between Egypt's struggle and their own.

This analysis poses a necessary question: do viewers genuinely see *The Square* as a starting point for their own social protests, or are they performing sympathy in public cyberspace? This question is both of deep concern to this essay and yet wholly rhetorical—it may be impossible to ever "know" the extent to which a political documentary inspires offline, "real" activism. The larger point may be that, regardless of any binary that exists between activism and slacktivism, these users who engage Netflix's review space find many kinds of value in *The Square*. For American reviewers, its lessons from "over there" are applicable to the situation "right here" across an unexpectedly wide array of political positions.

Conclusion: View, Share, Revolt, Rinse, Repeat

Issues of access, consumption, and shareability take on a different dimension when brought to bear on streaming subscription services such as Netflix. While important to recognize the technological and governmental barriers that exist in how users can engage with on-demand culture, this essay has placed the brunt of its analysis on the spatial dimensions of circulation. This negotiation between online and offline spaces, between civically-engaged media and civic participation, is ultimately a new dimension of Jane Gaines's "political mimesis." Gaines, in her discussion of political documentary, defines "political mimicry" as the generation of affect through "the conventionalized imagery of struggle."[73] The documentary form's indexicality, predicated on a relationship to Real spaces and Real people, "establishes a continuity between the world of the screen and the world of the audience, where the ideal viewer is poised to intervene in the world that so closely resembles the one on the screen."[74] This is again a version of spatial transgression—the space of representation carries into the lived space of the viewer, such that they are hopefully affected and compelled to "mimic" what the documentary shows them.

Especially as Netflix viewers have called *The Square* "a handbook" for revolution, it seems pertinent to consider Gaines's ideas in relation to media spreadability and circulation. "Freed" from the boundaries of the cinema screen, political documentaries like *The Square* become more mobile, but sequences of revolutionaries being spurred to action by social media videos also demonstrates a model of action. Apart from being a consciousness-raising project—documenting how the Egyptian Revolution happened from a number of individual perspectives—it also demonstrates how to use spreadable media to engender social change, while *itself* existing in a digitally distributed and shareable format. *The Square* is then a meta-text for shareable civic media, advocating both for social revolution and demonstrating how it has functioned in one specific site. This proposition begs a deeper understanding of what *The Square* can actually do in offline space. *The Square* may spur similar modes of film production, film acquisition, and film circulation. It may spur similar modes of civic engagement and challenges to current orders of government regimes in different geographical locations.

While Netflix's most notable non-series acquisitions of 2014 were a variety of comedy specials and a deal to produce several original feature films—including an exclusive deal on a series of Adam Sandler movies—it also released animal rights documentary *Virunga* in the latter half of the year, suggesting socio-political documentaries may remain a crucial—if limited or "boutique"—part of the service's distribution arm.[75] In that sense, Netflix uses its expanding collection of socio-political documentaries as a strategy to court one of its "highly differentiated micro-audiences."[76] Film festival acquisitions become, in this logic, another of Netflix's algorithmically-determined business practices designed to encourage certain types of viewers into subscriptions and habituated viewing practices more than a way of using their service to make a sort of political statement about the possibilities of digital film circulation.

The Square is part of the ongoing trajectory of participatory culture, spreadability, and civically engaged tactical media posing a challenge to what we *do* with increasingly mobile media, why we watch it, and why it is worth accessing. It represents a challenge to not just "rinse and repeat"—to binge-watch copious amounts of political documentaries through Netflix or other streaming services—but to engage the media we spread and circulate, and to think about ways to bridge online and offline spaces in new forms of civic engagement. The spreadability of digitally circulated political documentaries continues to expand our potential engagement with a plurality of civic spaces.

NOTES

1. Tim Markham, "Social Media, Protest Cultures and Political Subjectivities of the Arab Spring," *Media, Culture & Society* 36.1 (2014): 90.

2. Zeynep Tufekci and Christopher Wilson, "Social Media and the Decision to Participate in Political Protest: Observations from Tahrir Square," *Journal of Communication* 62 (2013): 366.

3. Neal Romanek, "YouTube Runs Netflix Oscar Nom The Square in Egypt," *TVB Europe*, March 4, 2014, accessed March 15, 2014, http://www.tvbeurope.com/youtube-runs-netflix-oscar-nom-the-square-in-egypt-2/.

4. W.J.T. Mitchell, *Cloning Terror: The War of Images, 9/11 to the Present* (Chicago: University of Chicago Press, 2011), 123.

5. As of this writing, the most recent information on network bandwidth comes from May 2014. See further: Todd Spangler, "Netflix Remains King of Bandwidth Usage, While YouTube Declines," *Variety*, May 14, 2014, accessed December 16, 2014, http://variety.com/2014/digital/news/netflix-youtube-bandwidth-usage-1201179643/.

6. Chuck Tryon, *On-Demand Culture: Digital Delivery and the Future of Movies* (New Brunswick: Rutgers University Press, 2013). Tryon charts a shift in cinema culture towards an increasing desire for instantaneity and evolving of flows of media through downstream windows.

7. Henry Jenkins, *Convergence Culture: Where Old and New Media Collide* (New York: New York University Press, 2006), 208.

8. See further: Douglas Kellner, *Media Spectacle and Insurrection, 2011: From the Arab Uprising to Occupy Everywhere* (New York: Bloomsbury, 2012); Manual Castells, *Networks of Outrage and Hope: Social Movements in the Internet Age* (Malden, MA: Polity Press, 2012); and Slavoj Žižek, *The Year of Dreaming Dangerously* (New York: Verso, 2012).

9. Mark Warschauer, *Technology and Social Inclusion: Rethinking the Digital Divide* (Cambridge: MIT Press, 2003), 38; emphasis in original.

10. Joel Penney and Caroline Dadas, "(Re)Tweeting in the Service of Protest: Digital Composition and Circulation in the Occupy Wall Street Movement," *New Media & Society* 16.1 (2014): 75.

11. Jan A.G.M. van Dijk, *The Deepening Divide* (Thousand Oaks, CA: Sage Publications, 2005), 173.

12. Manuel Castells, *Networks of Outrage and Hope*, 56.

13. Henri Lefebvre, *The Production of Space*, trans. Donald Nicholson-Smith (Malden, MA: Blackwell Publishing, 1984 [1974]), 26.

14. Castells, *Networks of Outrage and Hope*, 57.

15. Paolo Gerbaudo, *Tweets and the Streets: Social Media and Contemporary Activism* (London: Pluto Press, 2012), 11.

16. Gerbaudo, *Tweets and the Streets*, 5.

17. Nick Couldry and Anna McCarthy, "Introduction: Orientations: Mapping Media-Space," in *MediaSpace: Place, Scale and Culture in a Media Age*, ed. Nick Couldry and Anna McCarthy (New York: Routledge, 2004), 2.

18. Castells, *Networks of Outrage and Hope*, 10.

19. *Ibid.*, 57. Castells argues the burgeoning ordinariness of cell phone use in Egypt further contributed to this. This point certainly has resonances with Henry Jenkins's goal of

blurring the boundaries between producers and consumers/users, or the somewhat popular emergence of the word "produser," which collapses any sense of boundary between the two. See further Alex Burns, "Towards Produsage: Futures for User-Led Content Production," 2006, accessed December 15, 2014, http://eprints.qut.edu.au/4863/1/4863_1.pdf; and S. Elizabeth Bird "Are We All Produsers Now?" *Cultural Studies* 25.4–5 (2011): 502–516.

20. Castells, *Networks of Outrage and Hope*, 61.

21. Philip N. Howard and Malcolm R. Parks, "Social Media and Political Change: Capacity, Constraint, and Consequence," *Journal of Communication* 62 (2013): 360.

22. Zeynep Tufekci and Christopher Wilson, "Social Media," 370.

23. Merlyna Lim, "Clicks, Cabs, and Coffee Houses: Social Media and Oppositional Movements in Egypt, 2004–2011," *Journal of Communication* 62 (2013): 232.

24. This also has resonances with Lawrence Lessig's distinction between a "read only" culture of consumption, and a "read write" culture where users are actively encouraged to add to or otherwise change the media objects they encounter. See Lessig, *Remix: Making Art and Commerce Thrive in the Hybrid Economy* (London: Bloomsbury, 2008).

25. Henry Jenkins, Sam Ford, and Joshua Green, *Spreadable Media: Creating Value and Meaning in a Networked Culture* (New York: New York University Press, 2013), 2

26. *Ibid.*, 3, 6.

27. *Ibid.*, 7.

28. Carolin Gerlitz and Anne Helmond have developed similar ideas in their concept of the "Like economy," where "user interactions are instantly transformed into comparable forms of data and presented to other users in a way that generates more traffic and engagement" through a case study of Facebook's "like" and "share" buttons. See further Carolin Gerlitz and Anne Helmond, "The Like Economy: Social Buttons and the Data-Intensive Web," *New Media & Society* 15.8 (2013), 1349.

29. Joseph Turow, "Introduction: On Not Taking the Hyperlink for Granted," in *The Hyperlinked Society: Questioning Connections in the Digital Age*, ed. Joseph Turow and Lokman Tsui (Ann Arbor: The University of Michigan Press, 2008), 3.

30. Jussi Parikka, "Contagion and Repetition: On the Viral Logic of Network Culture," *Ephemera* 7.2 (2007): 288.

31. *Ibid.*, 289. See also Steven Shaviro, *Connected, or What It Means to Live in the Network Society* (Minneapolis: University of Minnesota Press, 2003).

32. Jenkins, Ford, and Green, *Spreadable Media*, 21.

33. *Ibid.*, 28.

34. Henri Lefebvre, *Everyday Life in the Modern World*, trans. Sacha Rabinovitch (New York: Harper and Row, 1987).

35. Jenkins, Ford, and Green, *Spreadable Media*, 39.

36. *Ibid.*, 155.

37. Ganaele Langlois, "Participatory Culture and the New Governance of Communication: The Paradox of Participatory Media," *Television & New Media* 14.2 (2012): 92.

38. See further Michel de Certeau, *The Practice of Everyday Life* (Berkeley and Los Angeles: University of California Press, 1984), 29–42.

39. Rita Raley, *Tactical Media* (Minneapolis: University of Minnesota Press, 2009), 3.

40. *Ibid.*, 6.

41. *Ibid.*, 10.

42. Michel de Certeau, *The Practice of Everyday Life*, 37.

43. Raley, *Tactical Media*, 14.

44. Evgeny Morozov, *The Net Delusion: The Dark Side of Internet Freedom* (New York: Public Affairs, 2012), 190.

45. *Ibid.*, 276.

46. *Ibid.*, 185.

47. Henry Jenkins and Nico Carpentier, "Theorizing Participatory Intensities: A Conversation about Participation and Politics," *Convergence: The International Journal of Research into New Media Technologies* 19.3 (2013): 266.

48. *Ibid.*, 281.

49. See further David MacDougall, *Transcultural Cinema* (Princeton: Princeton Uni-

versity Press, 1998), and Thomas Waugh, "Introduction: Why Documentaries Keep Trying to Change the World, or Why People Changing the World Keep Making Documentaries," in *Show Us Life: Toward a History and Aesthetics of the Committed Documentary*, ed. Thomas Waugh (Metuchen, NJ: Scarecrow Press, 1984), xi–xxvii.

50. Merrit Kennedy, "An Oscar Nominee, but Unwelcome at Home in Cairo," *NPR*, February 5, 2014, accessed March 15, 2014, http://www.npr.org/2014/02/05/271517965/an-oscar-nominee-but-unwelcome-at-home-in-cairo.

51. Jenkins, Ford, and Green, *Spreadable Media*, 219.

52. Chuck Tryon, *On-Demand Culture*. This stands somewhat at odds with the way Charles R. Acland has theorized the continued importance of cinema culture in the wake of downstream revenue windows. See Acland, *Screen Traffic: Movies, Multiplexes, and Global Culture* (Durham: Duke University Press, 2003).

53. Tryon, *On-Demand Culture*, 14.

54. *Ibid.*, 41.

55. Merrit Kennedy, "An Oscar Nominee, but Unwelcome at Home in Cairo"; Alex Ritman, "Jehane Noujaim's 'The Square' Makes Debut in Egypt," *The Hollywood Reporter*, June 7, 2014, accessed May 5, 2015, http://www.hollywoodreporter.com/news/jehane-noujaim-s-square-makes-710046.

56. Neal Romanek, "YouTube Runs Netflix Oscar Nom The Square in Egypt."

57. Blake Hallinan and Ted Striphas, "Recommended for You: The Netflix Prize and the Production of Algorithmic Culture," *New Media & Society* 18.1 (2014): 129; emphasis added.

58. Eli Pariser has called this kind of practice a "you loop," where the only content that gets directed at an online user is content that user has already expressed agreement or interest in. See Pariser, *The Filter Bubble: What the Internet Is Hiding from You* (New York: Penguin Press, 2011).

59. Thierry Kuntzel, "The Film Work," *Enclitic* 2.1 (1978): 38–61.

60. "Watch The Square Online," *Netflix*, accessed April 10, 2014. http://www.netflix.com/WiMovie/The_Square/70268449?

61. *Ibid.*

62. *Ibid.*

63. *Ibid.*

64. *Ibid.*

65. *Ibid.*

66. *Ibid.*; emphasis added.

67. *Ibid.*

68. *Ibid.*

69. *Ibid.*

70. *Ibid.*

71. *Ibid.*; emphasis in original.

72. *Ibid.*

73. Jane Gaines, "Political Mimesis," *Collecting Visible Evidence*, ed. Jane Gaines and Michael Renov (Minneapolis: University of Minnesota Press, 1999), 88.

74. Jane Gaines, "Political Mimesis," 92.

75. For more information on *Virunga*, see Steve Pond, "'Virunga' Is Netflix Documentary Gem: How Filmmakers Are Trying to Bring Down Big Oil," *The Wrap*, November 7, 2014, accessed Dec. 19, 2014, http://www.thewrap.com/virunga-is-netflix-documentary-gem-how-filmmakers-are-trying-to-bring-down-big-oil/. For more information about the Adam Sandler deal, see Todd Spangler, "Netflix Signs Adam Sandler to Exclusive Four-Movie Deal," *Variety*, October 1 2014, accessed December 19, 2014, http://variety.com/2014/digital/news/netflix-signs-adam-sandler-to-exclusive-four-movie-deal-1201319066/.

76. Hallinan and Striphas, "Recommended for You," 128.

Binge-Watching in Practice

The Rituals, Motives and Feelings
of Streaming Video Viewers

EMIL STEINER

Tomorrow: @HouseOfCards. No spoilers, please.—@Barack
Obama, February 13, 2014

Laurence Fishburne generally doesn't "do more than six episodes." Paul
Rudd "watched twelve episodes in a row." Judy Greer "watched the whole
season of *Veep* (2012–) in one sitting ... but that was during Sandy, the hurricane." Dennis Leary recommends doing it alone, but Michelle Monaghan
disagrees. Zosia Mamet says you can do it "in your birthday suit," but Keri
Russell prefers wearing "bad pajamas." Tom Riley favors "whiskey and a one-
sie," while Hannah New prefers "rum." Method Man likes "mangos," but for
Matthew Rhys it's "vodka and hot dogs."[1]
These are some of the descriptions of binge-watching practices and rituals that appear in a multiplatform advertisement from Comcast Xfinity
called "Celebrities Binge-Watch TV Too!" The commercial was posted to
YouTube on March 28, 2014, in the run-up to "Xfinity Watchathon Week," a
seven-day event each spring during which Comcast subscribers are given
free on-demand access to full seasons of popular programs from broadcast
networks and pay cable channels. The actors from some of those series appear
in the advertisement describing how they binge-watch.[2] The annual promotion is an attempt by America's largest cable provider to entice viewers into
subscription services by tapping into their voracious appetite for video content through binge-watching.
The informational and endorsing language of the promo spot indicates
that "binge-watching" is a nascent English term whose value is being negotiated in and through media. *Oxford Dictionaries* ranked "binge-watch" the

second most popular new word of 2013, behind only "selfie."[3] *Collins Dictionary* anointed it Word of the Year in 2015.[4] That same year a survey by Deloitte
found that "Two-thirds of viewers 'binge-watch' TV."[5] Apparently President
Barack Obama is one of them; so is Hillary Clinton.[6] As such, the liminal
linguistic position ("Has binge always been a verb?") and self-conscious usage
("but that was during Sandy, the hurricane") of binge-watching demand
deeper investigation.[7] Such celebrity endorsements may expedite acceptance
of an emerging behavior and serve as a model for its imitation.[8] However, is
Comcast's depiction representative of "binge viewers" generally? This essay
attempts to explore how non-celebrity audiences define, practice, and feel
about binge-watching and Netflix through qualitative, open-ended interviewing. My analysis of how viewers understand the behavior provides an orientation for academic and commercial research that points to a broader
reimagining of television's cultural identity. As one interviewee told me,
"Broadcast television is dead." The streaming video delivery popularized by
Netflix has empowered viewers to be more agentic consumers of culture,
while forcing traditional broadcasters and producers to adapt their content
and its delivery. That viewers have enthusiastically embraced the complicated
term binge-watching speaks to the complex, ambivalent, and ironic signification of post-industrial culture.

Theory Binge

During much of the twentieth century, scholars depicted the content of
broadcast television as mass-produced, low value entertainment and the audiences as gullible receivers staring glaze-eyed into flickering cathode-ray
tubes.[9] The Frankfurt School was particularly harsh in its deterministic critique of such popular culture; Theodor Adorno considered television a threat
to aesthetics with the potential for mind control. He and other Marxists theorists characterized television as a tool to perpetuate mindless capitalist consumption. Viewers, they argued, were trained into a culture of aspiration that
mitigated their ability and desire to question the political and economic structures of the powerful elite who controlled the airways.[10]

As the television set became a mainstream appliance in the 1950s and
1960s, American social scientists began empirically studying its "effects" on
audiences. Although audiences were not found to be automatons, effects
researchers did treat them as passive receivers.[11] Their findings indicated that
while the media cannot necessarily change minds, it can set the news agenda,
cultivate perceptions of reality, prime issues in viewers' minds, and influence
viewer behavior.[12] These examinations of media effects were largely unidirectional and ignored the individuality and agency of the viewer. Thus, from

not long after its mainstream inception, broadcast television's identity seemed fixed to the "Idiot Box" narrative—an industrialized system of commoditized, low-brow culture with the power to manipulate its passive viewers into consumerist ideology and potentially violent behavior.[13]

The Idiot Box narrative began to be challenged by cultural theorists of the 1960s and 1970s. During those decades, the "cultural sphere [was] divided into two hermetically separate regions."[14] Film, television, and popular literature (media) were studied through the theoretical lens of structuralism while "lived events" (rituals, customs) of the working-class were studied through "culturalism."[15] According to Tony Bennett, the former sought patterns in the content as evidence of the ruling class's domination of the masses through mechanized subservience. Conversely, the latter scoured popular culture for "romantic" and "authentic" expressions of subordinated voices shouting proudly against the tempest of domination.[16] Bennett proposed merging the two through Gramscian hegemony in order to "disqualif[y] the bipolar alternatives of structuralism and culturalism."[17] Such a move acknowledges the complexity of a viewer's power to negotiate meaning and the nuances of the economic and political structures behind television culturally and technologically.

Poststructural feminism scholars of the 1980s and 1990s rebranded audiences' voices as challengers to the status quo.[18] Building on Michel Foucault, Chris Weedon examined experience through subject position and motive, activating the audience's role in discourse.[19] Janice Radway encouraged negotiation and renegotiation of the "nature of the relationship between audiences and texts."[20] Meanwhile, Jacqueline Bobo challenged the Frankfurt School's top-down power dynamics and the unidirectional flows of early social scientists: "Producers of mainstream media products are not aligned in a conspiracy against an audience."[21] Recognizing the relationship of producers and viewers as cooperative can "legitimize the audiences' scrutinizing gaze," while complicating traditional power dynamics.[22]

The Idiot Box narrative has been further complicated in recent years by the technological blurring of what television is, the ambivalence associated with digital labor, "performances and the production of user-generated content," and self-conscious audience participation.[23] In this post-industrial narrative, the viewer and the broadcaster share and negotiate the meaning of texts. Series that failed on broadcast television, such as *Arrested Development* on Fox (2003–2006; 2013), have been reincarnated on Netflix through viewer demand and the affordances of subscription based streaming video. Even popular series, like *Breaking Bad* (2008–2013), saw their cultural footprint expand further as a result of Netflix releasing prior seasons so new fans could catch up through binge. As one interviewee told me, "*Unbreakable Kimmy Schmidt* (2015–) is made for Netflix, not NBC," the network for which the

Tina Fey–produced sitcom was originally created. Producers use viewer feedback to shape scripts, while viewer behaviors like binge-watching have encouraged writers to craft complex narrative arcs designed to be viewed in bulk rather than once a week.

Building on this theoretical history, I decided to examine how audiences binge-watch from a perspective that both acknowledges the structure and profit motive of media companies like Netflix that encourage binge-watching, while also recognizing the agency of audiences to demand and control how and what they watch. Netflix has branded itself as a catalyst for the changes in broadcast power structures and information flow, but it is most certainly a for-profit company that benefits from the rise of binge-watching and the collection of viewers' personal preferences, interests, and data.[24] At the same time, viewers exploit Netflix to inexpensively carry and access troves of content on their terms, and perhaps influence what content is produced.[25] The unidirectional broadcaster-to-viewer transmission model of the twentieth century is now a dynamic circulation of content, responses, and meanings. I hoped to reflect that dynamism in my audience interviews. My goal was to locate, inhabit, and reflect binge-viewers' "structure of feeling" within and without the nostalgia of television's Idiot Box narrative and Netflix's profit motives.[26] As Katherine Sender points out, "Participants' research reflexivity offers a frame to reconsider contemporary debates about audience research and the role reflexivity might play in these debates."[27]

The nascence of binge-watching poses challenges for traditional understandings of television and for traditional methods of research. Historically most communication scholars have treated the research of new technology as science and examined user behaviors with quantitative, socio-psychological tools.[28] I believe that such research can be valuable for close-ended research questions and for developing large samples with predictive and organizational potential. Because binge-watching is a hybrid of technology and culture, I also believe that a wide range of methods should be used to analyze it.[29] However, to unearth how and why people binge-watch, I chose to start with qualitative interviewing with open-ended questions that acknowledge the audience's active role in the meaning making of culture.[30]

My approach favors openness over objectivity. By openness I mean accepting that my consciousness is in constant discourse with every aspect of this project. Rather than attempt to isolate, contain, or control that subjectivity, I decided to acknowledge, encourage, and embrace it. My goal was for that openness to be reflected by the interviewees so that the conversations became dialogues of equal positioning.[31] This made me a better listener and, ideally, created an atmosphere for freer exchange. The objective was subjectification. I will therefore unpack my own history, behavior, and motives for binge-watching as my interviewees did for me.

The first time I read the term binge-watching was in the December 2011 issue of *Wired*, but I had been doing it for years.[32] The first series I can remember binge-watching was *The Sopranos* (1999–2007). I had resisted the HBO drama at first, but, in August 2002, I had some downtime and decided to give it a try. With the fourth season scheduled to air in September 2002, I rented—from a video store—the DVDs of season one. Over the next week, I watched all 39 episodes from the first three seasons. I watched alone, on a 32-inch CRT, in stretches of no less than three episodes. Because I was on vacation, my viewing times and habits were erratic. I remember watching until late at night and then waking up and wanting to watch more. I was binge-watching, I just did not know to call it that. Had I asked myself, as I have the interviewees in this study, why I was viewing so voraciously, I would have said that the series was too good to stop. As soon as an episode ended I wanted more, and, unlike with traditional broadcasting, I could have more—as much more as I wanted. I also felt like I finally understood why so many people loved the series, and I wanted to catch up so I could be a part of the conversation.

After some 40 hours, I could not wait until the September premiere of season four, and I did not. I started watching the first three seasons again. It was around then that my mother told me I was watching too much television. She was right, and I felt guilty, ashamed, lethargic, but I did not stop. Over the next 13 years, I watched more series than I can remember, but I do remember most of the series that I binge-watched. Like most people today, binge-watching is my preferred method of consumption, but I am aware of its perceived effects; it is a pleasure whose control I am constantly negotiating. Now I study binge-watching, and I am aware that my own rituals and motives influence how I perceive the behavior in myself and others. I believe that that influence can provide insight on shared user experiences, but it may also cause me to overlook details that would stand out to a non-bingeing researcher. My choices of questions and my analysis of viewer responses are unique to me, and I acknowledge that unique position.

A Brief History of Binge Technology

DVDs make for inefficient binge-watching. Despite the tactile aesthetics of a DVD box set, viewers using that medium have to insert a disc every three or four hours, which involves an interruption of the narrative immersion.[33] In 2002 I did not have much choice, but that soon changed. On-demand programming and DVRs made it possible for viewers to watch continuously without having to get out of their seats. If VHS and DVD were the first generation binge-watching technology, DVRs and on-demand services were the second-generation. Soft-launched in the Bay Area in 1998, TiVo was one of the first DVRs to compress and save video from television to a hard drive. Its popularity grew in the early 2000s as a device that allowed viewers to pause live television and return to it later. It also allowed viewers to record programs and watch them later without commercials (by fast-forwarding), like a VCR minus the VHS tape.[34] In order for a series to be saved on TiVo, or any other DVR, it would first have to air on television and be recorded by the viewer. On-demand programming eliminated the necessity to record and store content, but it also limited choice. A viewer could only order what his or her cable provider offered on-demand.[35]

In the first years of the twenty-first century, the "Not TV" programming narrative touted by HBO increased the demand for "highbrow" content while streaming video was simultaneously changing the model for content distribution.[36] A perceived improvement in programming, particularly serialized dramas, spiked in the mid–2000s with series like *24* (2001–2010, 2014), *The Wire* (2002–2008), *Six Feet Under* (2001–2005), and *Deadwood* (2004–2006).[37] The perceived improvement along with increased Internet bandwidth and HDTV pushed DVD viewing out of vogue.[38] The void was filled by what I consider to be the third generation of binge-watching technology: digital media players (DMPs) and later smart televisions. These set-top boxes connect online content to televisions, virtually eliminating the need for on-demand.[39] In cooperation with streaming services like Netflix, Hulu, HBO Go, and Comcast's Xfinity Go, which allow users to watch a variety of content on computers and mobile devices, DMPs like Apple TV, Roku, and Amazon Fire TV, seamlessly merged those services on large HDTV screens.[40] Yet, it was Netflix's full-season release model, which began with the Norwegian co-production *Lilyhammer* (2012–2014), that finally set the table for binge-watching to become the all you-can-eat buffet for which viewers had been salivating.[41]

From a content perspective, Netflix's syndication of programming helped popularize series that had faded, or even completely failed, under the traditional broadcast model.[42] Netflix's vast programming budget sparked a demand for original content production that has created the current programming arms race among traditional television networks and digital content providers. The resulting glut has swelled the potential size of personal video libraries beyond the viewing capacity of a human lifetime.

Binge-Defining

Los Angeles Times culture critic Mary McNamara provided one of the first formal definitions of binge-watching: in January of 2012: "Binge television: n. any instance in which more than three episodes of an hourlong drama or six episodes of a half-hour comedy are consumed at one sitting. Syn.: Marathon television and being a TV critic."[43] Over the next two years, Google searches for "binge-watching" spiked.[44] Since the publication of McNamara's article, Netflix's stock price has soared over 700 percent. Binge-watching is now considered "the new normal," but the behavior's signification remains ambiguous.[45] Is binge-watching revolutionary, dangerous, manipulative, empowering, or all of the above?

Until 2012, the noun binge connoted unhealthy behavior—a period of uncontrollable excess. As a verb it is still commonly associated with binge drinking and binge eating—psychological symptoms associated with a pathological loss of control.[46] Despite appending the same modifier, binge-watching has mostly been depicted as a liberating experience, the worst side-effect of which is poor personal hygiene, in the more than ten thousand newspaper and magazine articles mentioning the behavior since 2012.[47] The rapid transformation of binge-watching from obscurity to ubiquity has stretched popular understanding of binge. The act and the term represent a subversive use of a signifier to ironically exaggerate the signified. The wordplay stems from the first people to identify as *binge viewers*—self-conscious 1990s "TV nerds" who passionately celebrated their viewing obsessions with like-minded outsider fanatics.[48] The "nerd-only" behavior has gone mainstream without losing its ironic vestigial root modifier. As one interviewee told me, "Binging [*sic*] is cool because it's still subversive." Of course, Netflix has been quick to capitalize on that "cool," while simultaneously making fun of itself and binge-watchers.[49] Despite all of the chatter about binge-watching, few media scholars have studied the behavior. In late 2013, Harris-Interactive and Netflix conducted proprietary research through an online survey of over 3,000 U.S. Netflix customers, of which nearly 1,500 streamed series at least once a week. Of them, 73 percent reported "positive feelings towards binge-streaming TV" and 80 percent said "they would rather stream a good TV show than read a friend's social media posts."[50]

Binge Method

As I have argued, binge-watching is a hybrid of technology and culture. It challenges the traditional power dynamics of unidirectional broadcast flows from producer through viewer. To reflect this, I employed qualitative methods and grounded theory in an attempt to put the viewer as interviewee on a more equal footing and to reflect the dynamism of the behavior itself. I conducted this exploratory research on binge-watching during 2014 through semi-structured interviews each lasting approximately 60 minutes as well as informal discussions each lasting approximately 30 minutes. My rationale was that in-depth conversations provide more substantive and robust answers to how and why questions, and elicit the kind of thick descriptions that can encourage future scholarship. The conversations were conducted in-person. Twenty-one women and 15 men between the ages of 22 and 66 participated. They all lived in the Philadelphia or New York City metropolitan areas, and most had completed a college degree; four were born outside the United States. I reconnected with several participants during the spring of 2015 to ask follow-up questions via Facebook Messenger.

My interviewing style was strongly influenced by my eight years as a journalist. I used open-ended questions and a semi-structured format that allowed our conversations to flow organically. Although most of the people I spoke with answered (directly and indirectly) my 25 questions/prompts, they did not all answer them in the same order or by my prompting. An interviewee may have answered question 12 ("Can you describe your typical binge-watching experience?") when I asked question seven ("What are some of the things you enjoy about binge-watching?"). Again, my goal was to loosen the interview structure in order to liberate rich, personal insights rather than sticking to a script.

After completing and transcribing the initial interviews, I began to look for themes. I employed qualitative, inductive methods to code the content.[51] As the interview process evolved, I noted when those themes arose in other conversations. As I found repetition in the themes, I sorted for categories of rituals and motivations while continuing to have discussions that influenced the sorting. This dynamic process continued until late 2014, when I reached a saturation point; I kept finding the same themes arising from different questions. I then felt satisfied that I had gathered sufficient information to conclude the interviews.

Next I coded the transcripts and employed theoretical sampling of the thick descriptions I had culled in order to isolate motives, rituals, and feelings.[52] This afforded me a vivid tableau of how this group of people define binge-watching, how and why they binge-watch, and how they feel about binge-watching. I should note that participants were very eager to discuss binge-watching with me. This led to snowball sampling: interviewees told people they knew about my project, allowing me to expand my network of participants. The popularity and novelty of the behavior are certainly responsible for participant enthusiasm, but I believe that people find the ambivalence of binge-watching intriguing as well. As Mary Choi writes in *Wired*:

> Weird stuff happens after about eight hours of watching the same TV show. Your eyes feel crunchy. You get a headache that sits in your teeth, the kind that comes from hitching your free time to a runaway train of self-indulgence—too much booze, food, or sleep. Of course, there's also a sense of accomplishment, of smugness, that comes from blowing through years of television in mere days.[53]

That ambivalence speaks to how binge-viewers are actively negotiating their behavior. It is an ironic balancing act of feigned remorse and vacillating pride/shame. It is both empowering and debilitating—an experience of control and lack of control facilitated by technology. I felt that tension of comprehension and signification vibrating through our conversations.

Binge Findings

Rituals

Interviewees roughly affirmed McNamara's 2012 definition of binge-watching—at least two hours of the same 30-minute series or at least three hours of the same 60-minute series. An important difference was that some interviews considered binge-watching to be simply watching a series "from beginning to end." While most interviewees stressed that the number of episodes watched in a sitting was primary to the definition, there were at least two interviewees for whom the consistency and completion of a season or series were more important. A 55-year-old English teacher from Philadelphia told me that she might only watch two episodes of a series per night. "As long as I'm consistently watching that show, then yeah, it's binge-watching." She only had time to watch short spurts, but she perceived a season as a complete unit that she was finishing "like a book," on her own schedule. Her definition of binge-watching was about controlling the narrative flow and closure at her convenience rather than the slower broadcast model she had grown up using. Those who still had cable television (about half) reported bingeing through on-demand platforms as well as streaming video. Others used only streaming devices; no one regularly used DVDs, though many interviewees' first bingeing experiences were, like mine, with DVD players or VCRs. Predictably, Netflix was the most commonly cited service for binge-watching across all devices.

Interviewees differentiated binge-watching from other television watching in terms of portability and consistency. While some interviewees acknowledged that it is possible to binge-watch broadcast commercial television, such as a *Seinfeld* (1989–1998) marathon, and many had at some point watched "traditional" television for extended periods of time, no one reported binge-watching through that medium. Several interviewees told me: "I hate commercials" often with a colorful modifier inserted. A 29-year-old South Jersey graduate student told me, "I find traditional TV annoying now. Even DVR. I don't like commercials. I don't like how commercials get super loud. I don't like waiting for the next episode. Netflix makes TV better." Advertising, particularly when it is obvious and intrusive, interrupts the focused continuity and narrative immersion that viewers associate with binge-watching.

The distinction of the Netflix user experience is such that some younger viewers perceive the service as other than television, even if they watch Netflix on a television. Viewers' ability to watch on multiple devices (smartphones, laptops, tablets), their technological control of content, described as being able to pause, rewind, and fast-forward, and the full season release model were identified as essential to the experience. They all challenge the traditional broadcast model of commercial breaks. "When I need to go to the bathroom I can pause," one person told me. "I don't need to wait for a commercial." A 22-year-old engineering student reported that she stopped watching the MTV comedy *Awkward.* (2011–2016) because she found the commercials "annoying." Before Netflix she may have stuck with the series; now she can resist it.

According to my interviews, most binge-watching takes place at home, in the evenings of workdays, and on weekends. A 31-year-old software engineer did admit binge-watching at work, but that was only because he was "being laid off over the course of six months" and "had to be in the office" even though he "wasn't really working." Interviewees reported using vacation time to binge series that they didn't have time to watch earlier. "I do plan on clearing my schedule for a few days for *OINTB* (*Orange Is the New Black* [2013–])," a 30-year-old writer from Philadelphia told me. "I have taken sick days to finish a show," one interviewee admitted sheepishly. "Long weekends usually mean Netflix," said another. Viewers celebrate the freedom of being able to control the content consumption while acknowledging the power of the content to control them.[54] As one interviewee told me in August of 2014, "I'm not going near *Game of Thrones* (2011–) until Christmas." Although season four of the HBO fantasy drama premiered in late spring of that year, he chose to savor the epic by waiting for long holiday when he could be completely immersed. This was his choice, but he also recognized that it might be dangerous for him to start the series while he was working.

Binge-watching is most often a solitary behavior, especially when series are viewed on mobile or handheld devices. "It's hard to share an iPad for three hours," one interviewee pointed out. Several interviewees mentioned trying to binge with their spouses; one had binged with her roommate who didn't like TV: "I got her into *Vampire Diaries* (2009–2017)." An interviewee described the experience of bingeing *Lost* (2004–2010) with his wife through Netflix as cooperative and trust-based. "If I watched ahead that would be cheating." When I asked if he had ever cheated on his wife, he admitted he was tempted, especially when she was out of town.

Due to conflicting schedules, tastes, and energy levels, collaborative binges are typically shorter than solitary binges. Binge-watching groups of greater than two are irregular and far less common according to interviewees, though one recalled a group marathon in college of *Law & Order* (1990–

2010). Another interviewee had attended a *House of Cards* (2013–) party when Netflix released the second season in February 2014. "But no one stayed for all 13 episodes." The logistical constraints of sharing consumptive control make "group bingeing" less common than solitary bingeing. However, many interviewees reported binge-watching alone so that they could discuss the series with other people, or be part of a perceived ongoing cultural conversation. One interviewee liked to text her friends while they binged the same series separately. "We can't be together for *Orange Is the New Black*, but we'll text each other."

Interviewees described binge-watching along a continuum of attentive to inattentive, which I coded as the Viewer Attentiveness Spectrum (VAS). More attentive bingeing is a focused study of the text that is both entertaining and educational, often motivated by the need to catch up or feel narratively immersed. Less attentive bingeing is almost always for relaxation, nostalgia, and distraction. The level of attentiveness is a product of the content, but interviewees determined which content they watched based on how attentive they wanted to be. Sixty-minute serial dramas like *Mad Men* (2007–2015) and *Homeland* (2011–) are series that demand attentiveness. Episodes often end with cliffhangers that entice viewers to stay tuned. During more attentive bingeing, the goal is to actively absorb, analyze, and be immersed in the content, which may be narratively complex and emotionally taxing. Some interviewees noted that they rewind and re-watch scenes to improve their understanding of the plot, characters, and dialogue. "Like *House of Cards,* if I was getting up to drink something, I would hit pause because the text is complicated." The complication is part of the entertainment and the allure to keep watching in high VAS bingeing.

Viewers described the content that they binge-watch less attentively as "background noise," which "doesn't take a lot less effort" to watch. Although they may have the series on for several hours, they reported doing other activities such as "folding laundry," "cooking dinner," and "grading papers" while the episodes played. Interviewees named sitcoms with single-episode plots like *The Office* (2005–2013) and *Parks and Recreation* (2009–2015), procedural dramas like *Law and Order* and *House* (2005–2012), and reality shows with formulaic structures and frequent recaps like *19 Kids and Counting* (2008–2015) and *Keeping Up with the Kardashians* (2007–) as the series they binged less attentively. Interviewees seemed aware of a connection between VAS and content. Many respondents reserved attentive bingeing for longer periods of downtime, like vacation; less attentive bingeing was for any time they wanted to "relax and just have something on." Re-bingeing a series also tends to be lower VAS, particularly for comedies.

A few interviewees referred to highbrow versus lowbrow content as a factor in the type of binge-watching they did, but most stated that the difference in VAS was determined by the structure and complexity of the series. "If you're watching an episode of *Lost,*" said one Philadelphia graduate student, "and there's a sandwich in one scene, you need to know where that sandwich is, or you miss something." Noticing such details may be crucial in a mystery; in a sitcom they are usually less relevant to the experience. "I don't pause or rewind," said an interviewee of her less attentive bingeing of *Arrow* (2012–) on Netflix. "I could be in the kitchen making coffee, then five minutes later I come back and [the Arrow] is still beating people up. I don't feel the need to stop everything and obsessively watch something like that. It takes a lot less effort to watch a series like *Arrow* than a series like *House of Cards* or *Game of Thrones,* which would be one I would pause."

Viewers found advertising more intrusive for series that they were watching attentively. A clever viewer described his strategy of "DVRing" the first 20 or so minutes of *Mad Men* so that he could then begin the episode and fast-forward through commercials as the series aired. This sensitivity to advertising extends to product placement. "If Don Draper [the protagonist of AMC's *Mad Men,* played by Jon Hamm] is drinking Coke during a meeting that makes sense," the viewer told me. "But if Ned Stark [Sean Bean's character on *Game of Thrones*] had a Coke in his hand, it just wouldn't make sense in that world." The viewer's attention would be called to this inconsistency, which would pull him out of the story. At the same time a series that demands greater attentiveness, like *Game of Thrones,* "is just better when you binge-watch it." Comments like these indicate that narrative immersion and attentiveness work symbiotically to enhance the experience of binge-watching while simultaneously being enhanced by the act of binge-watching. Through their narrative form, lower VAS series tend to be less immersive, which affords more mental energy for multitasking as well as less hostility toward commercial interruption.

Bingeing Motives

Everyone I spoke with reported at least three of the following reasons for binge-watching: (1) enhanced viewing experience; (2) sense of completion; (3) cultural inclusion; (4) convenience; (5) catching up; and (6) relaxation/nostalgia. Almost all interviewees stated that watching an entire season at once was more pleasurable than having to wait a week between episodes, though one self-described obsessive viewer said she was glad she had to wait for some series because "I can't control myself." Interviewees also preferred the Netflix model of full season releases to the traditional one episode per week broadcast model. "Even if I don't watch the whole season, I like to have the option." The perceived authenticity of this enhanced viewing experience was a common theme I found. An interviewee told me that bingeing is how a series "should be watched" because it allowed him to "get inside the writer's head." If each season of a series is written as a unified arc, then binge-watching allows viewers to experience the arc without interruption. This perception of scriptwriting appears to reinforce the value of binge-watching over the traditional once-per-week tune in.

Interviewees also felt that the variety and quality of content had improved with binge-watching. "There are so many shows out there that are so good," one interviewee said proudly. "You read more articles about award show snubs than about the shows that win the awards." I coded an underlying sentiment that binge-viewers bore some responsibility for making programming "more intelligent" and prolific over the past five years. This jibes with the Netflix narrative that binge-watching empowers viewers by giving them control over programming.[55] Interviewees cited the ability to create lists and to rate series as empowering, though no one stated that those ratings directly changed programming. Instead, the perception was that producers were responding to the increased attention afforded through the technology. "There's just so much goodness," said one respondent with the wistful esurience of an epicure at an endless feast. While the empowerment narrative of bingeing was subtly laced through my interview transcripts, viewers openly cited their ability to communicate instantly about series with other viewers around the world as evidence of their power to indirectly affect programming. Viewers rarely expressed concerns about providers surveilling their viewing habits—in fact, some seemed heartened that their preferences could be heard and potentially affect content.

Several interviewees said that being able to finish a series immediately was a motivation for binge-watching. The convenience of completion speaks to the control that binge-watching technology affords viewers: "If I want to watch two whole seasons of *Friday Night Lights* (2006–2011) in a weekend I can ... and I have." Another told me, "I hate waiting a week to watch the next episode." Being able to watch whenever you like was a benefit often cited. "A lot of times I'm at the office until nine at night, so I can't tune in like my parent's generation," said the 30-year-old writer. The convenience of the technology also allows viewers to be picky. "I don't have to watch a show when it comes on because it's only on then. I can wait to read reviews or see what my friends say about it." As a result viewers may watch more, but that they may also watch more selectively.

The ability to participate in a series' discourse community motivates some of the longest binges. Interviewees were bashful about admitting that they had watched hours of programming in a short period of time so that they could "fit in" with friends, colleagues, and strangers, though as they opened up during our conversations, I observed cultural inclusion to be a consistent motivator of binge-watching. One interviewee who works at a large, public university described a group of colleagues whom she respects discussing *Orange Is the New Black*. "I wanted to be part of the cool club," she admitted. This kind of discourse group also serves to filter viewing selections. Because the interviewee admired her colleagues, she perceived their discussion as evidence of the series' culturally relevance.

Less professional discourse groups provide different motivations for binge-watching. Two interviewees reported binge-watching *House of Cards* because they knew friends would be talking about it on social media, and they did not want the surprises spoiled. "I'd have to avoid human contact," one joked. The cultural inclusion motivator extended to online communities. Some expressed a guilty sense of pride at what posting about a series signified. To announce you have Netflix or HBO is to make an announcement of class. A 37-year-old investor considered himself "ontologically incompatible" with anyone who likes *Friends* (1994–2004). "We can work together, but we can never understand each other."

Viewers' inclusion in online discussions can also motivate them to binge-watch, but it carries a burden of obsessiveness. "I used to participate in fan communities," one interviewee told me ruefully. "It was such a time suck." She quit when she found a job. Respondents also noted changes in the viewing discourse communities afforded by technology. "We don't have the water cooler conversations anymore," a 54-year-old media professor told me. "Because of Netflix we just don't have to wait until Friday to talk about Must See TV [from Thursday night]." Another interviewee believed there was more talk. "The way we watch is really communal right now," she said referring to the online conversations that take place through social media. Viewers appreciate the communal and empowering aspects of these communities, but they also recognize the extra time participation requires in addition to the hours they already spent watching. "I had to quit," said the interviewee who found a job. "It was too much."

There were three people who described themselves as compulsive readers and compulsive watchers. A 28-year-old real estate developer told me.

> I'm either going to hate it or I'm going to like it, and if I like it I'm going to watch all the episodes in a week. I do the same thing with books. Once I start a book it's very difficult for me to put it down. I will sit there for 18 hours with a book. I don't know if that's related or my personality, but that's the way I am…. There's this feeling inside me, and I want all my questions answered … and it's on Netflix so I can keep going…. I get a sense of relief when I finish. I have to finish.

That same interviewee also admitted that her need to complete a series was "a problem." For these people the sense of completion was the most powerful motivation for binge-watching. "I can go on until there's no more show…. I've never met anyone who consumes media the way I do." One interviewee stated that he refused to start a series until he knew it had ended. "I need to know that I can complete it." The need for completion often led to the longest binges and was often associated with negative feelings. Netflix's Post-Play function, which starts the next episode in a series automatically, makes stopping harder. While the function improves narrative immersion, it may lead to more compulsive viewing than traditional television or even DVDs where a viewer has to get up and insert a new disc. Even viewers who had attentively binged series that they thought of as highbrow sometimes felt they had "overbinged." Others felt this was a ridiculous notion. "I only feel bad about binge-watching if the show sucks," the software engineer quipped. At the same time he also mentioned feeling anxious about all the series he "has to watch," and he stated that he didn't binge-watch during vacations. For him binge-watching was "like a job," albeit one that he enjoyed.

The Structure of Binge Feelings

My conversations indicate that viewers have an ambivalent relationship with binge-watching. Interviewees described the variety of viewing platforms and the breadth and quality of content as "amazing" and "overwhelming." They described the viewing controls as "convenient," and the ability to watch "whatever, whenever, without commercials" as empowering. But interviewees also expressed regret at the compulsiveness the controls and ability could cause. "Netflix is the devil!" one interviewee joked in reference to the company's Post-Play function. "You have to tell Netflix not to play the next episode.... You could be dead, and the episodes would keep playing." At the same time, interviewees felt entertained by and excited about binge-watching. Many felt that producers now cater to binger-viewers with better series than traditional broadcast television. "I think the way writers are writing now is different," the real estate developer told me. "It is accommodating to a smarter audience—one that has access to every other viewer and episode online." These observations confirm some the findings of the PwC survey of new television habits.[56]

About half of the interviewees noted similarities with binge-watching and reading. The South Jersey graduate student argued that binge-watching *House of Cards* was more intellectual than reading *Fifty Shades of Grey* (2011). "Why isn't that binge-reading?" she demanded. The convergence of books and video onto portable technology appears to be blurring traditional distinctions between book reading and television watching, while the perceptions of the quality and cultural position of television are rising. This may contribute to the sense of pride or reduced shame of high VAS bingeing, particularly of "highbrow" content. "I can take my Netflix library on vacation.... [I]t's my beach reading," a Philadelphia-area graduate student boasted.

Interviewees also used terms typically associated with addiction (compulsion, withdrawal, overdose, functional binger, etc.) to describe their binge-watching habits. Some reported feelings of regret and self-loathing after longer binges. The engineering student told me that binge-watching when she has a hangover makes her feel "like a loser." No one reported losing a job or a relationship because of bingeing, although several interviewees admitted being less productive because of long binges. However, their admissions were often tempered with pride, particularly after binges of "highbrow" content. "I'm such a nerd," said one interviewee while proudly describing how he had "re-binged" *Mad Men* on Netflix. Terms associated with endurance sports such as "hitting the wall," "second wind," and "commitment" were also used to describe binge-watching. The connotations associated with *marathon* differ from those of *binge*, though some viewers used them interchangeably for television.[57] One interviewee corrected me at the start of the interview: "I hate the term binge-watching, I prefer the term marathoning … binge-watching sounds like something guilty…. Bingeing is never seen as something healthy." Despite Pheidippides's fate, marathons are seen as healthier than binges.

Interviewees characterized low VAS bingeing as a worse use of their time than more attentive bingeing. Satisfaction derived from the relaxation/ nostalgia motive was described as short-lived, particularly with reality shows.[58] "I'm not going to sit there and marathon [*Here Comes*] *Honey Boo-Boo* (2013–2014)." However, some intentionally avoided series that they felt they would need to watch attentively because of the time commitment. "I'm afraid to get into a series like *Doctor Who* (1963–1989, 1996, 2005–); that would be two months of my life." That same interviewee—a mathematics doctoral student in Philadelphia—was comfortable bingeing episodes of the animated comedy *Family Guy* (1999–2001, 2005–) because he could stop watching it more easily and had seen most of the episodes before. He was a compulsive reader motivated to binge-watch by the sense of completion. The longest and most obsessive binge-watching is usually more attentive, though these are also the binges of which people seem most proud. This may be related to the higher cultural status of series that require attentive bingeing. The closer to literature the programming appears to be, the healthier its obsessive consumption appears to feel.

Ambivalent Medium

As cultural anthropologist Grant McCracken points out, television is no longer the "vast wasteland" described by Federal Communications Commission (FCC) Chairman Newton Minow in 1960, but an engrossing and rewarding cultural space: "TV has changed and we are changing with it."[59] The Idiot Box narrative though is difficult to shake, but here's a suggestion: embrace it. If binge-watching is an escape from reality, so be it. After all, how focused is our multi-tasking reality today? When a binge-viewer attends one series for four hours, it may be his/her most focused activity all day. When s/he binge-watches to relax, s/he can now do so with more control and variety than in the homogenous Camel Caravan era.[60] When you binge-watch you do not channel surf, you do not have to watch commercials, and you can move about the series, and the world, freely. You make a conscious choice to closely read a text or to relax with soothing background sounds and old favorites. High VAS binge-watching is akin to reading a book in structure and practice, if not in medium. Serialized dramas are divided into consecutive episodes, often called chapters, while their content is consumed sequentially like a novel.[61] Chapters often conclude with an unresolved conflict to stoke curiosity. Cliffhangers have been used for decades in television, but today's technology and the Netflix delivery structure have empowered viewers to scratch that itch now, like with a book, rather than waiting a week for the next thrilling installment.

Perhaps the medium is perception, but the medium is changing. You can now carry a library of books and series on the same device and consume as much as you like. But television's passive/addictive narrative is still connoted in the word binge.[62] Imagine you spent the weekend reading a great book. You stayed up late turning pages and woke up early excited to jump back in. Maybe you finished the book in one epic sitting, or you devoured it in several bites. Would you feel ashamed of dedicating your whole weekend to reading? Would you feel embarrassed to tell colleagues or friends? Should spending your weekend engrossed in a television series be different?

Culturally the difference is perceptual: books are traditionally associated with eggheads, television with couch potatoes. Although media scholars and journalists have long insisted otherwise, my interviews suggest that people still perceive reading books as a culturally and intellectually superior activity to watching television.[63] The terrain of television's cultural value is being rearticulated, but it is a slow process. Younger viewers tend to differentiate between Netflix binge-watching and television technology, while older viewers do not. There were some interesting contradictions during the interviews, for instance this exchange with the South Jersey graduate student:

RESPONDENT: TV with commercials is so boring. I have neglected TV.
E.S.: You don't consider Netflix to be TV?
RESPONDENT: Oh. I do.… Actually maybe I don't. Haha! Sorry that's not helpful.
E.S.: No, it is.
RESPONDENT: I think of TV as traditional, turn on the TV when a favorite show is on or maybe channel surfing. My TV watching is much more purposeful now. I don't channel surf as much. TV is more an active activity for me now rather than passive? If that makes sense?

Netflix has a lot to do with the shifting perception, but aspects of the old stigma remain and are evident in the guilt with which interviewees perceive their binge-watching. As one compulsive viewer admitted to me at the end of the interview, "I'm going to have a lot to think about as I reevaluate my behavior." Although she chuckled at the statement, it was clear she felt guilty about how much television she watched. The stigma is rooted in popular perceptions of television consumption. If television's content is cultural fast food then its consumer must be a tasteless, morbidly obese glutton for the lowest common denominator. But as John Fiske points out, "The lowest common denominator may be a useful concept in arithmetic, but in the study of popularity its only possible value is to expose the prejudices of those who use it."[64] Netflix is rebranding television as a gourmet meal, and the viewer an epicure. Media corporations may be getting richer the more that idea is swallowed, but so too are the content and audience experiences. But gluttony can happen in a three-star restaurant where the chefs still "worship at the altar of audience ratings."[65] If you believe that binge-watching has liberated audiences, you must appreciate the irony that the expression of that freedom is massive, obsessive consumption. The struggle for control of meaning making is bound by the liminality of the binge-viewer's identity and his/her practice of binge-watching.

This convergence of technology and culture is complicated and indeed seismic.[66] Binary theoretical oppositions are insufficient to unpack it. To appreciate its nuances we must understand that binge-watching cannot be either positive or negative, cultural or structural, but an evolving human experience driven and energized by contradiction. Through this lens, a binge-viewer can be a bookworm and a couch potato. Binge-watching can be an addictive behavior and a meditative one. That is the ambivalence, rooted in the contradiction of pleasure negotiations that these conversations demonstrate.[67] As media scholars analyzing binge-watching and Netflix, we should consider what C.W. Mills called the "sociological imagination"—a perspective that allows the observer to occupy multiple subject positions simultaneously.[68] Doing so promotes a mirroring of subject and object: the (ad)vantage of seeing the forest and the trees.

Notes

1. Xfinity, "Celebrities Binge-Watch TV Too," accessed March 28, 2014, https://www.youtube.com/watch?v=ARulpPItWRs.
2. *Ibid.* The spot also aired on a Comcast's on-demand main menu during the same period.
3. "Oxford Dictionaries Word of the Year 2013," accessed February 24, 2016, http://blog.oxforddictionaries.com/press-releases/oxford-dictionaries-word-of-the-year-2013/.
4. Collins Language, "'Binge-Watch'—Collins Word of the Year 2015," *Collins Dictionary,* accessed February 24, 2016, https://www.collinsdictionary.com/word-lovers-blog/new/binge-watch-collins-word-of-the-year-2015,251,HCB.html.

5. Deloitte, "Digital Democracy Survey: A Multi-Generational View of Consumer Technology, Media and Telecom Trends," *Digital Democracy Survey, 9th Edition.* (2015), 11, accessed February 24, 2016, http://www2.deloitte.com/us/en/pages/technology-media-and-telecommunications/articles/digital-democracy-survey-generational-media-consumption-trends.html.

6. HUFFPOST TV, "Which TV Shows Does Obama Watch?" *The Huffington Post*, December 30, 2013, accessed February 24, 2016, http://www.huffingtonpost.com/2013/12/30/obama-tv_n_4518832.html; "Hillary Clinton Binge-Watches *The Good Wife*," *The Late Show with Stephen Colbert*, accessed May 1, 2016, https://www.youtube.com/watch?v=_by4NUtNARY.

7. Xfinity, "Celebrities Binge-Watch TV Too!"

8. Amishi Arora and Khushbu Sahu, "Celebrity Endorsement and its Effect on Consumer Behavior," *International Journal of Retailing & Rural Business Perspectives* 3.2 (2014): 866–869.

9. Theodor Adorno, "How to Look at Television," in *The Culture Industry*, ed. Jay M. Bernstein (New York: Routledge, 2001), 158.

10. *Ibid.*, 160. See also Max Horkheimer and Theodor Adorno, "The Culture Industry: Enlightenment as Mass Deception," in *Mass Communication and Society*, ed. James Curran, Michael Gurevitch, and Janet Woollacott (Beverley Hills: Sage, 1977), 349–383.

11. Elihu Katz, "The End of Television?" *The Annals of the American Academy of Political and Social Science* 625.1 (2009): 6–18.

12. Paul Lazarsfeld, Bernard Berelson, and Hazel Gaudet, *The People's Choice* (New York: Columbia University Press, 1944); George Gerbner, "Toward "Cultural Indicators": The Analysis of Mass Mediated Message Systems," *AV Communication Review* 17 (1969): 137; Joan D. Schleuder, Alice V. White, and Glen T. Cameron, "Priming Effects of Television News Bumpers and Teasers on Attention and Memory," *Journal of Broadcasting & Electronic Media* 37.4 (1993): 437; Albert Bandura, Dorothea Ross, and Sheila A. Ross, "Imitation of Film-Mediated Aggressive Models," *Journal of Abnormal and Social Psychology* 66 (1963): 3.

13. Jason Mittell, "The Cultural Power of an Anti-Television Metaphor: Questioning the 'Plug-in Drug' and a TV-Free America," *Television & New Media* 1.2 (2000): 215–238; Laurie Ouellette and Justin Lewis, "Moving Beyond the 'Vast Wasteland': Cultural Policy and Television in the United States," *Television & New Media* 1.1 (2000): 95–115; Harold Mendelsohn, "Socio-Psychological Construction and the Mass Communication Effects Dialectic," *Communication Research* 16.6 (1989): 813–823; Katz does point out that early "theorizing by David Sarnoff (1941) … went on to predict that the new medium would bring people "home," integrate the nation, and raise cultural standards, while also warning against the potential of political "showmanship," the power of audiovisual advertising, and the danger of ideological propaganda." See Katz, "The End of Television," 8.

14. Tony Bennett, "Popular Culture and the 'Turn to Gramsci,'" in *Cultural Theory and Popular Culture: A Reader*, 4th edition, ed. John Storey (Harlow, UK: Pearson Education Limited, 2009), 82.

15. *Ibid.*, 84.

16. *Ibid.*, 86.

17. *Ibid.*, 83.

18. Chris Weedon, "Feminist Practice and Poststructuralist Theory," in *Cultural Theory and Popular Culture: A Reader*, 4th edition, ed. John Storey (Harlow, UK: Pearson Education Limited, 2009), 320.

19. David Morley, *Television, Audiences, and Cultural Studies* (New York: Taylor and Francis, 1992), 18.

20. Janice Radway, "Reading *Reading the Romance*," in *Cultural Theory and Popular Culture: A Reader* 4th edition, ed. John Storey (Harlow, UK: Pearson Education Limited, 2009), 199.

21. Jacqueline Bobo, "The Color Purple: Black Women as Cultural Readers," in *Cultural Theory and Popular Culture: A Reader* 4th edition, ed. John Storey (Harlow, UK: Pearson Education Limited, 2009), 367.

22. Katherine Sender, *The Makeover: Reality Television and Reflexive Audiences* (New York: New York University Press, 2012), 2.

23. For instance, see: Ethan Thompson and Jason Mittell, *How to Watch Television* (New York: New York University, 2013), 2; Mark Andrejevic, "Watching Television Without Pity: The Productivity of Online Fans," *Television & New Media* 9.1 (2008): 24; and Sarah Banet-Weiser, *Authentic™: The Politics of Ambivalence in a Brand Culture* (New York: New York University Press, 2012), 64; Sender, *The Makeover*, 24.

24. Mareike Jenner, "Is this TVIV? On Netflix, TVIII and Binge-Watching," *New Media & Society* (2014): 3.

25. Ally, "Binge-Watching Is the New Normal," *Fandom Obsessed*, January 20, 2014, accessed February 24, 2016, http://fandomobsessed.com/binge-watching-is-the-new-normal/.

26. Raymond Williams, *The Country and the City* (New York: Oxford University Press, 1973), 58.

27. Sender, *The Makeover*, 165.

28. James Carey, *Communication as Culture, Revised Edition* (New York: Routledge, 2008), 12.

29. Emil Steiner, "Binge-Watching Framed: Textual and Content Analyses of the Media Coverage and Rebranding of Habitual Video Consumption" (unpublished manuscript, Temple University, 2014).

30. Ien Ang, *Desperately Seeking the Audience* (New York: Routledge, 1991), 68.

31. Amanda D. Lotz, "Assessing Qualitative Television Audience Research: Incorporating Feminist and Anthropological Theoretical Innovation," *Communication Theory* 10.4 (2000): 447–467.

32. Mary H.K. Choi, "In Praise of Binge TV Consumption," *Wired*, December 27, 2011, accessed February 24, 2016, http://www.wired.com/2011/12/pl_column_tvseries/.

33. Jason Mittell, "Serial Boxes," *Just TV*, accessed February 24, 2016, https://justtv.word press.com/2010/01/20/serial-boxes/.

34. For an examination of VCR and cable TV's effect on network viewing see Dean M. Krugman and Roland T. Rust, "The Impact of Cable and VCR Penetration on Network Viewing: Assessing the Decade," *Journal of Advertising Research* 33.1 (1993): 67–73.

35. For cultural impacts of digital delivery see Chuck Tryon, *On-Demand Culture: Digital Delivery and the Future of Movies* (New Brunswick: Rutgers University Press, 2013).

36. Marc Leverette, Brian L. Ott, and Cara Louise Buckley, ed., *It's Not TV: Watching HBO in the Post-Television Era* (New York: Routledge, 2008), 37. See also: Thompson and Mittell, *How to Watch Television*, 312.

37. Leverette, Ott, and Buckley, *It's Not TV*, 2.

38. Sam Laird, "How Streaming Video Is Killing the DVD," *Mashable*, 2012, accessed February 24, 2016, http://mashable.com/2012/04/20/streaming-video-dvd-infographic/.

39. The first "set-top boxes" were actually video game consoles, though these lacked synchronicity with robust digital content services initially.

40. Saurabh Goel, "Cloud-Based Mobile Video Streaming Techniques," *International Journal of Wireless & Mobile Networks* 5.1 (2013): 85–93.

41. A viewer can watch, as I did, the first four episodes of *House of Cards* at home and then pause in the middle of the fifth episode to be dragged to the mall, but continue watching that episode on his/her cell phone in the car and then catch the sixth episode on an iPad at the Apple Store.

42. Zac Stockton, "Netflix Won't Own Binge-Viewing for Much Longer," *reels*, July 21, 2014, accessed February 24, 2016, http://www.reelseo.com/netflix-wont-own-binge-viewing/.

43. Mary McNamara, "Critic's Notebook: The Side Effects of Binge Television," *Los Angeles Times*, January 15, 2012, 1.

44. Google Searches for: "binge-watching," Worldwide, *Google Trends* (2004-present), accessed February 24, 2016, http://www.google.com/trends/explore#q=binge%20watching& cmpt=q.

45. "Netflix Declares Binge Watching Is the New Normal," *PRNewswire*, December 13, 2013.

46. "Oxford Dictionaries Word of the Year 2013."

47. Brian Stelter, "New Way to Deliver a Drama: All 13 Episodes in One Sitting," *New York Times*, February 1, 2013, sec. A; Business/Financial Desk; Steiner, "Binge-Watching Framed," 4.

48. "Oxford Dictionaries Word of the Year 2013."

49. Isabella Biedenharn, "Netflix's Binge-Watching Warnings are the Best April Fool's Prank," *Entertainment Weekly*, April 1, 2014, accessed February 24, 2016, http://www.ew.com/article/2015/04/01/netflix-binge-watching-psa.

50. "Netflix Declares."

51. Kathy Charmaz, *Constructing Grounded Theory, 2nd Edition* (Los Angeles: Sage, 2014), 109–224.

52. *Ibid.*

53. Choi, "In Praise of Binge TV Consumption."

54. Despite the claims made in "Celebrities," food and eating was not claimed to be an integral part of binge-watching rituals by my interviewees.

55. Ally, "Binge-Watching Is the New Normal"; Marcus Wohlsen, "When TV Is Obsolete, TV Shows Will Enter Their Real Golden Era," *Wired*, May 15, 2014.

56. PwC, "Feeling the Effects of the Videoquake," 6.

57. In "Binge-Watching Framed," I argue that the distinction between marathon watching and binge-watching is in the consistency of content. A movie marathon, for instance, may be many different movies, but binge-watching is always the same show.

58. For more on reality TV audience satisfaction and motives, see Lisa R. Godlewski and Elizabeth M. Perse, "Audience Activity and Reality Television: Identification, Online Activity, and Satisfaction," *Communication Quarterly* 58.2 (2010): 148–169.

59. Grant McCracken, "5 Things You Don't Know about Binge Viewing" (unpublished manuscript, Harvard University, 2013), 1.

60. Patrick Barwise and Andrew Ehrenberg, *Television and Its Audience* (London: Sage, 1988), 127.

61. This format may have started with DVDs where the digital technology was better suited to jumping between sections than the smooth rewind and fast-forward of mechanized VCRs.

62. Ian Christie, *Audiences: Defining and Researching Screen Entertainment Reception* (Amsterdam: Amsterdam University Press, 2013), 159. For a deeper discussion of passive/addictive narrative see Seth Finn, "Television Addiction?" An Evaluation of Four Competing Media-Use Models, *Journalism Quarterly* 69.2 (1992): 422–435; Mittell, "The Cultural Power of an Anti-Television Metaphor."

63. Jenner, "Is this TVIV," 5.

64. John Fiske, *Television Culture* (New York: Methuen, 1987), 309.

65. Pierre Bourdieu, "Television," *European Review* 9.3 (2001): 251.

66. For more on the integration of audience in and through new television technology see Elizabeth Evans, *Transmedia Television: Audiences, New Media, and Daily Life* (New York: Routledge, 2011); Christine Quail, "Television Goes Online: Myths and Realities in the Contemporary Context," *Global Media Journal* 12.20 (2012). For more on digital discrimination see Joseph Turow, *Niche Envy: Marketing Discrimination in the Digital Age* (Cambridge: MIT Press, 2006); Steinar Ellingsen, "Seismic Shifts: Platforms, Content Creators, and Spreadable Media" *Media International Australia, Incorporating Culture & Policy* 150 (2014): 106–113.

67. Storey, *Cultural Theory and Popular Culture*, 98.

68. C. Wright Mills, "The Sociological Imagination," in *Social Theory: The Multicultural and Classic Readings*, ed. Charles Lemert (Boulder: Westview, 1999), 351–352.

Narrowcasting, Millennials and the Personalization of Genre in Digital Media

ALISON N. NOVAK

In 2006, Netflix announced a $1 million prize for research teams who could successfully help them improve their rating algorithm. The goal was to use the large amounts of data collected from users to make better film and television suggestions for individual users.[1] By 2012, researchers dedicated over 2,000 hours to combine 107 algorithms that would produce automated recommendations for users. Netflix branded this addition to the platform as a feature of personalization, control, and convenience, and remarked on its innovation in the online streaming world.[2] Regarding Netflix's ongoing push towards improving the automated system, scholars noted that while the improvements were innovative, the practice of making suggestions to viewers was already in place as a far-reaching trend known as "narrowcasting."[3]

This essay explores the drive towards narrowcasting and Netflix's practice of making personalized recommendations for users based on its rating system. While narrowcasting has been an industry norm for years, its oppositional relationship to the historical model of broadcasting suggests that Netflix's personalization is a force of change within the film and television industry. Today, many other platforms share Netflix's narrowcasting model and make personalized recommendations for users through collected data. Therefore, this essay looks at how narrowcasting is presented to users, and how the practice is embraced or understood—both important and timely issues due to the rising popularity of streaming media.

Importantly, this study considers these reactions while looking at a specific audience of Netflix: 13- to 33-year-old users, also commonly known as "millennials." Previous research has indicated that millennials are the largest and most coveted demographic group for media producers. However, relatively little is known about their relationship to narrowcasting, and previous research has made predictions regarding their ability (or lack thereof) to be critical to media platforms and suggested content. Ultimately then, this study seeks to add to our knowledge of this rising group and their relationship to a growing media trend.

Netflix, Algorithms and Data Mining

Years after the 2006 Netflix Prize, Vice President of Product Innovation and Personalization Algorithms, Carlos Gomez-Uribe, revealed that the company employs more than 800 engineers responsible for developing recommendation algorithms and maintaining the personalization of the site.[4] At the heart of the operation are data algorithms, which analyze user behavior in an effort to make specific content recommendations. This requires that user actions are recorded and then analyzed. Netflix Engineer Director Xavier Amatriain states:

> We know what you played, searched for, or rated, as well as the time, date, and device. We even track user interactions such as browsing or scrolling behavior. All that data is fed into several algorithms, each optimized for a different purpose. In a broad sense, most of our algorithms are based on the assumption that similar viewing patterns represent similar user tastes. We can use the behavior of similar users to infer your preferences.[5]

While Netflix uses individualized datasets to make recommendations, these are often based of larger trends among its 40 million users. Most of the data collected is used primarily for internal purposes to deliver personalized content to each account; however, the large amount of data also reveals overall trends. For example, Netflix engineers learned that while personal ratings of each movie are important, they are more likely aspirational rather than reflective "of daily activity."[6] Therefore, the majority of recommendations are based on actual content viewed and the searching/browsing patterns of users, rather than what is explicitly noted by individuals. This is reinforced by research on "algorithmic culture" that argues algorithms and other digital information archiving is more indicative of users' true identity than what they articulate on digital media.[7] The larger practice of collecting user information is known as "data mining," a term used by social and information scientists in their practice of identifying, collecting, and then analyzing big data.[8] The desire for data mining develops from demands to thoroughly investigate how users interact with new technology, media, and digital platforms that are now commonplace in the twenty-first century. As such, data mining has received much attention and debate surrounding its use and role in user privacy.[9]

This debate shows up frequently in conversations surrounding Netflix. Without a doubt, Netflix has revolutionized how site data is collected and analyzed. The Netflix Prize was aimed at doing just that. However, this collection of data is also viewed by some as an invasion of privacy, or as a way to hide the recording of specific user behaviors.[10] Despite requiring all users to agree to an End User License Agreement spelling out the company's collection and recording rights, a growing group of critics argues that these practices are problematic and potentially harmful to both users and the greater media industry. Felix Salmon of *Reuters* suggests that while Netflix states this data collection is used for making personalized recommendations, the actual content provided on the instant-streaming platform challenges this belief.

> The original Netflix prediction algorithm—the one which guessed how much you'd like a movie based on your ratings of other movies—was an amazing piece of computer technology, precisely because it managed to find things you didn't know that you'd love. More than once I would order a movie based on a high predicted rating, and despite the fact that I would never normally think to watch it—and every time it turned out to be great. The next generation of Netflix personalization, by contrast, ratchets the sophistication down a few dozen notches: at this point, it's just saying "well, you watched one of these Period Pieces About Royalty Based on Real Life, here's a bunch more." Netflix, then, no longer wants to show me the things I want to watch, and it doesn't even particularly want to show me the stuff I didn't know I'd love. Instead, it just wants to feed me more and more and more of the same, drawing mainly from a library of second-tier movies and TV shows, and actually making it surprisingly hard to discover the highest-quality content.[11]

Salmon indicates that Netflix's data mining is falsely described as making the user experience better, when in reality it actually hinders the user's ability to see desired content. These concerns echo similar debates surrounding the emerging trend of narrowcasting.

Narrowcasting

In the early 1960s and 1970s, those studying television and the growing cable industry spoke frequently about the potential for content specialization and personalization of the television experience for every audience.[12] Although still early in its theoretical foundation, narrowcasting began to take shape as an ideal and goal for future cable businesses that sought to deliver individualized content to small segments of the public. Whereas early business models of cable desired to deliver generalized content to large segments of the population, the drive towards maximizing profits lead business leaders to dream of creating specialized series, movies, and television experiences for each audience member. Despite this optimism for the future, narrowcasting as a practice developed in film and television in the 1980s, decades after the original vision.[13]

Narrowcasting is defined as the process of reaching audiences by identifying facets of users' identity to target programing and genres. This is the process of creating or displaying content in a way that targets small, specific portions of the audience. Throughout the late 1980s and 1990s, networks and advertisers turned their attention to producing specific televised content for targeted audiences. Economically, the goal was to identify a small subset of the larger audience, research and build an understanding of their needs and interests, and then create content specifically for those groups. In return, these groups were thought to be more loyal, more attentive and more willing to share messages from those series and media producers. Narrowcasted content was levied as a way to forge stronger bonds between media producers and small subsets of the population, thus encouraging the perception of a one-to-one relationship emerging within the mass medium.

Furthermore, early proponents of narrowcasting argued that this trend could assist producers in developing programing that showcased underrepresented or marginalized groups.[14] Beretta E. Smith-Shomade notes that narrowcasting was "heralded as a viable prescription for global understanding," a path for producers to include previously taboo or unpopular topics and genres for smaller, accepting audiences.[15] Likewise, Susan Tyler Eastman, Sydney W. Head, and Lewis Klein add that narrowcasting emerged as a viable and economically appealing trend because the practice acknowledged that some facets of the population were currently underrepresented or left out of mass-appeal broadcasts.[16] Narrowcasting reinforces previously successful economic models of identifying unmet needs of a population and subsequently creating and tailoring services to that group.[17] As a result, narrowcasting is boasted as a democratic part of the future of media, one where more groups are represented, and more space is created for a variety of audiences (including those traditionally left out or ignored).[18]

However, narrowcasting has also been critiqued for its inability to live up to the democratic promises first associated with its integration into media production models. Despite its ability to recognize and build programing for parts of the population who are traditionally left out, researchers argue that the audience is still spoken of as a homogenous group, especially considering how ratings are calculated and advertising dollars are spent.[19] The priority, particularly when it comes to television, is still to generate small numbers of programs that target large facets of the population, rather than many programs that target small facets of the population.[20] As a result, narrowcasting's presence in traditional media formats, including television, radio, and film has been subject to critique from academics and industry professionals. Many acknowledge that although tailoring media content to specific segments of the population is a regular practice, the democratic potential of narrowcasting has not yet been met.[21]

However, narrowcasting's appeal is not limited to traditional media formats; today's digital media is filled with examples of platforms, websites, and services designed for and dedicated to small, specific audiences. Peter Ludes notes that in digital technologies and media, narrowcasting is often associated with audience fragmentation, where audiences are broken up into smaller subsets.[22] This results in tailored advertising campaigns to those small audiences, thus making the small groups more valuable than large, diverse ones. Although strikingly similar to narrowcasting in description, researchers have identified two ways that fragmentation differs. First, narrowcasting is related to content creation for entertainment and informational purposes. Fragmentation relates to advertising or narrowing down audiences for delivery to an outside content producer (generally in marketing).[23] Second, fragmentation is the result of identifying small subsets of the population, while narrowcasting is the process of separating audiences from each other.[24] While each practice might reinforce or affect the other, they are viewed as two distinct practices.

A frequent area of inquiry in narrowcasting in digital media has been looking at how audiences are identified and targeted in political campaigning.[25] These studies emphasize that online platforms including social media, tracking software, and electronic communication enable major digital organizations to record substantial information about users, and sell large datasets to advertisers, who then can create marketing materials (e.g., ads and popups) tailored to specific groups identified as possible supporters.[26] Digitally, narrowcasting has been identified as a global paradigm shift in economics and production.[27] Five of six of the largest, most successful international media conglomerates have adopted some type of narrowcasting strategy in their creation of digital content, including Internet and mobile television.[28] This includes the creation of specialized websites, advertisements, and videos available to audiences based on their user profiles and other collected information from their digital experiences. Helen Wood notes that the prevalence of narrowcasting in digital content has created a demand for new ways of identifying and measuring audiences.[29]

This has also resulted in increased revenue for digital media providers who implement narrowcasting into their platform. TechWeb argues that revenues from narrowcasting increased 90 percent from 2005 to 2009, and are continuing to rise each year.[30] Advertisers are willing to pay for their content to be delivered to smaller audiences because the rate of success is higher. Similarly, platforms offering narrowcasting platforms such as Netflix and YouTube have seen their own profits increase because of the convenience offered to potential users. As a result, WirelessNews.org named digital narrowcasting its 2007 trend to watch, recognizing that companies that branded themselves as narrowcasting and personalizing content for users were going to become more profitable and popular within the next ten years.[31]

Millennials and Media

Among the groups narrowcasting sought to identify and create content for, millennials and American youth emerged as a targeted demographic. Although scholars debate the classification of millennials as under-represented in the mass media, without a doubt, new narrowcasting platforms and content such as Netflix sought to attract the coveted demographic. Jennifer Gillan notes that American youth are now used to content and recommendations made based on their demographic data, as this trend emerged long before online and digital streaming.[32] American youth growing up in the 1990s found that content was regularly narrowcasted towards them, meaning created to engage and attract the group in an effort to gain popularity and advertising dollars.

The millennial generation was born between 1981 and 2001, and is some-times called "Generation Y" or "Digital Natives." One of the largest genera-tional groups in history, the demographic has not escaped academic or mainstream criticism for its relationship to media and digital platforms. A frequent assessment of the group is that its expectation of narrowcasting and targeted media content has produced a narcissistic or self-absorbed genera-tion.[33] *Marketing Weekly News* reported that millennials expect their digital platforms to present recommended content (including news, advertising, and entertainment) that fit with their interests and personality. Narrowcasting is no longer just a technique used to make the platform stand out; it is now an expected part of the media experience for millennials. This in turn, means millennials expect to see programing and recommended titles that reflect their own lives, thus producing a narcissistic label.

Similarly, other research has identified millennials as lacking critical media skills necessary to help them understand long term consequences of framing, agenda setting, and even narrowcasting. D.T.Z. Mindich suggests that as millennials engage with less traditional sources of news (newspaper, radio, and television) and increasingly rely upon new media (Internet-enabled technologies), they become complacent (in comparison to other generational groups) and accepting of content presented to them.[34] Rather than seek out information, they are too reliant upon media producers tailoring content to their interests, therefore making them less likely to think negatively or even critically about media practices.

However, other scholars suggest that millennials have found new ways to articulate their uncertainty about the changing media landscape. Although they still rely upon these platforms to provide tailored content, when qualitatively interviewed about their media consumption, they reflect discourses of active disengagement and an understanding of how to critically engage with the content presented to them. For example, Debora S. Vidali notes that millennials are keenly aware that tailored news and platforms present personalized media while leaving out other content deemed not of interest or not likely to be engaged with.[35] This awareness manifests when millennials discuss their reactions to media platforms, but is likely ignored in quantitative surveys designed to learn about millennial media habits. As a result, this essay will use qualitative journals as a means of understanding how millennials engage with narrowcasted content and the Netflix platform. In addition to exploring millennials reactions, the study will look for evidence of millennial critique or predictions of long-term consequences.

Methods

To investigate how millennials interacted with Netflix and understood narrowcasting, a journal analysis was completed. Previous research has identified journals as being a successful tool for learning about millennials media habits.[36] Nick Couldry, Sonia Livingstone, and Tim Markham have previously examined how journals are useful in understanding the engagement of a user with information media.[37] Journals are a critical tool to understand user experiences because they allow for immediate recording of reactions to media content, rather than waiting for an appointment with an interviewer. This decreases the risk of forgetting emotions or changing reactions to fit with an interviewers behavior. Similarly, Laura Harvey used journals to qualitatively study how individuals reacted to legal information.[38] Journals are also important in recognizing the contextual experience of media viewing. Niall Bolger, Angela Davis, and Eshkol Rafaeli suggest that journals allow users to describe the contextual experience of a media interaction, such as what else they are thinking about, working on, or paying attention to during a media engagement.[39] Ultimately, journals allow users to document experiences, examples, and past stories that relate to the topic at hand.

Respondents were selected for the project based on their enrollment in a general education course at a large northeastern university. The project was offered as an extra credit assignment for students who have a Netflix account and an alternative assignment was provided for students who opted not to participate. In total, 27 millennial-age students participated and submitted journal entries on their experiences with Netflix. Students were asked to write a one-time journal entry as a response to a Netflix themed prompt. The prompt asked participants to login into their Netflix account and look at recommended titles, content, and categories. Then, they were asked to reflect on these recommendations and the knowledge they had of Netflix's practices. No prior instruction on Netflix was given to participants, meaning their reflections and information provided was based on individual knowledge and experience.

Specifically, this study looked at how students explained Netflix's practices and business model. The goal was to identify what information millennials had about the narrowcasting in the platform and what they thought the implications that practice would have on technological development and other facets of society. By providing an open-ended prompt, participants were encouraged to elaborate in a variety of ways including providing examples, making comparisons, or reflecting on long-term effects. As a result, the findings in this study provide insight into millennials understanding and reaction to narrowcasting and the Netflix platform.

Findings

Millennial journal entries revealed that the group is largely critical of the recommendations made through the Netflix platform, as well of the process and data collected by the company to develop targeted programing. Three topics were addressed by the entries and are explored below. These include: (1) identifying Netflix's method; (2) reflecting on the user experience; and (3) making future predictions about the role of narrowcasting in media.

Finding One: Method Identification

Of the 27 millennials who submitted journal entries on Netflix, 20 identified, explained, and analyzed the methods used by company to tailor recommendations and programing. Frequently, these explanations appeared early in the writing samples, and were used as a means of entry to talk about larger issues of privacy, ease-of-access, and data mining.

Most millennials identified algorithms and computer learning strategies as Netflix's method to making film recommendations. For example, a 20-year-old male said:

Netflix uses algorithms and metadata to provide recommendations for users. These professionals are gaining intimate knowledge of your interests, emotions, and beliefs in order to gain a perspective of how you are so that the artificial intelligence can make a profile and try to curtail advertisements and certain recommendations based solely of this profile. A perfect example of this type of intelligence is Netflix.

These posts strictly identify Netflix as a company seeking to learn more information about the user so that they may more finely make recommendations and maintain user loyalty and attention. Importantly, these reflections are mostly descriptive in nature, rather than critical or persuasive. This description is a technique used to introduce the topic and then later as evidence.

There were also posts that delved deeper into the types of data collected by Netflix and the extent of their data mining. Participants clearly identified the outward limits of the company's ability to target, monitor, and collect information about users. Again, these are descriptive reflections rather than a persuasive tone. Three other 20- to 25-year-old respondents said:

> Netflix digs deep and tracks: When you pause, stop, fast forward or rewind, the date and time you watch content, what zip code you are in, ratings, browsing and scrolling, searches and when you leave content and if come back. Netflix basically tracks every movement. A good thing about this analytic is that Netflix will know what shows are more likely to be canceled, or help deal with ending credits.

> Platforms such as Netflix are able to learn about your interests because they monitor what exactly you search, and exactly what movies you watch. After they gather this information, they analyze it in a database, and everything else that you do in your account, and then they are able to guess what movies they think that you would want to watch.

> Netflix learns about your interest by analyzing the different genres, titles, and actor based off of you history on the platform. While some of the recommendations are good, there are also some that are completely off.

The two posts above particularly emphasize the participants' focus on the personalization and extent to which user data is documented and analyzed to provide insight into customer preferences. The terms "monitor" and "tracks" suggest that participants understand the practices that are used regularly by Netflix to tailor content to the user.

There were also posts that provided examples of the Netflix process in action. These examples demonstrate how aware participants are of the methods and uses of the data provided. Here, participants started to connect narrowcasting to their own lives, particularly emphasizing the recommendations made for their own accounts. The introduction of the subjective "I" as well as sharing of personal experiences suggests that millennial participants are not only aware of the general practices of Netflix; they also understand how it influences their own lives. Three other participants added:

> For example, I watched a Disney movie and after I was done, Netflix recommended similar Disney movies.

> The more someone uses Netflix, the more the Netflix computer learns about the user. For example, if someone seems to be watching a lot of comedies, Netflix retains this information and suggests more comedies that it thinks the user may be interested in, and the process repeats.

> Like millions of other Americans, Netflix is an important part of my life. While

there are still some shows that I will watch on an actual television, most of my TV watching occurs on Netflix. The more I use Netflix, the more it learns about me as a person. While this sounds like the worst nightmare for Dwight Schrute [Rainn Wilson] from *The Office* [2005–2013], it can still be beneficial to the user. Netflix learns about my viewing interest through several ways. First, is me watching shows or movies. If Netflix, sees that I am re watching *The Office* for the fourth time, it may recommend me to watch *Parks and Recreation* [2009–2015], which I am currently re watching for the second time, as that too is a quirky NBC sitcom. Second, Netflix has a rating system for its programming and I can give shows a rating from 1 to 5 stars. Netflix can use this and recommend me shows that are similar to the ones I give four or five stars. Third, you can also tell Netflix the type of genres you prefer. I like comedies, but I am not generally a fan of anime shows. Thus, Netflix will not recommend me something like that.

These posts also describe how the relevant content worked within the context of their lives. The participants give Netflix agency in their description of the processes used to analyze user behaviors. Through phrases such as "Netflix sees" or "Netflix computer does," the participants document the work being done by the company to be a part of their user experience. This agency will be discussed in the coming sections.

However, some participants noted their confusion over the Netflix process and method. Seven of the 27 journals noted they were familiar with the general process, but still confused as to the specifics of how the system worked. Again, these entries connected the practices of the organization to the user experience.

I am not sure exactly how it works, but I assume there are little people inside of the TV writing down notes (haha). The notes that these elves take aren't always the best, because sometimes Netflix go off in a whole different direction regarding programs that its viewer would want to watch.

In more appropriate terms though, Netflix offers relevant recommendations because it uses machine learning to create and compare the data of watch and search history for all of its users.

The reflection on the process reveals that although contributors were never asked to address or describe how Netflix's platform works, users are still aware and informed about narrowcasting's presence and the methods used to achieve it. Overall, most of the participants reflected descriptively on the processes utilized by Netflix to tailor content selections. Importantly, most participant journals started with this description, which was then used as a point of entry to critique or analyze the success of these recommendations, which is described in the next section.

Finding Two: User Experiences

Similar to other studies on millennials and the media, journal entries in this study revealed that the group viewed narrowcasting in both positive and negative terms. However, these reflections were almost entirely focused on the user experience rather than larger implications of these trends.

FAVORABLE ANALYSIS

When discussing the positive traits of narrowcasting in Netflix, many identified recommendations as helpful features of the site, as they introduced users to new content or previously unknown films and television series. Two 18-year-old men and one 21-year-old woman added:

> This is really helpful seeing as though the Netflix library is so large to sift through the whole catalog to find a program to watch would take time that our fast paced short attention spanned generation would not like to waste just looking for something to watch. This technique is very good for business. There is also a list of titles dedicated solely to the programs that our social media friends have been watching.
>   This is great considering that the best form of advertising for centuries had been word of mouth and personal testimonials. How great would it be to go to Netflix not knowing what to watch and seeing a list of things that your friends are watching? This creates intrigue and not only creates a larger list of recommendations for yourself, but also gives you something to talk about the next day with your friends.
>   Netflix and computer learning in general can help people to discover things about themselves that they would probably never learn all on their own. Sometimes you need someone else to pick out your flaws to fix them, and then sometimes you need someone to pick movies and spam them along the "Recommended for [Name]" tab, the latter is Netflix. I can't remember exactly what Netflix says about their computer learning, other than how simple recommendations make watching movies and shows. There's not much more other than the facts I was dishing out earlier in this blog post. Netflix allows a lot of possibility for consumer to reach into the metaphorical bag and pull out a handful of excitement.

Again, the emphasis here is on the user, and how the system benefits the individual. Rather than looking at how this may influence broader society, eight of the millennial participants in this study reflected on how narrowcasting influenced them or the people in their immediate surroundings. As two 18- to 25-year-olds noted:

> For me, these recommendations do fit my interest because it gives me options that I would like to see that I didn't think of searching for. However, it also gives me new options that I never seen, but I usually have no interest in. I would say the recommendations have a 70 percent success rate on my use of the platform.
>   The benefits with this form of machine learning greatly outweigh the negative aspects. With the learning of your viewing preferences, Netflix basically takes the hassle out of search for something to watch. This type of computer learning is sold by providing a free trial for 30 days with upon completion requires a monthly subscription to continue usage.

These eight reflections ultimately focused on how the recommendations affected them. This is demonstrated through terms such as "I have" or "my interests." These terms of reference are important as they denote who or what the millennial participants view as most important or critical to the Netflix platform. Previous research suggests that millennials often view their media use as individually, rather than socially impactful. However, while there is some evidence of this, there are also an equal number of journal entries that suggest greater society may also be impacted. These entries suggested that narrowcasting allowed users to access content that they may have otherwise ignored, thus exposing them to new points of view. On a large scale, this practice would help everyone become aware or have the opportunity to be aware of diverse points of view. As one 23-year-old participant put it:

> When I use Netflix, I do feel that the recommendations are fitting to my interests. I have found some of my favorite shows and series through the recommendations that Netflix suggests. I think the idea of computer learning is a good thing for society. On a positive note, computer learning makes it much easier for the user to find content that they already have a strong interest in.

Importantly, posts like the one above suggest that millennials view media effects as important to both the individual user and larger society. This is an important finding, as it provides insight into how millennials view the role and impact Netflix has both on the media industry and culture.

Negative Analysis

Alternatively, 18 entries discussed the negative features of Netflix and the narrowcasting process. First, many added that although the system was based on learning from your own, recorded behavior, the recommendations made were far from perfect. Thus, recommendations were a distraction or took up more time from the user as they attempted to wade through the imperfect fits. Sample responses in this category include:

> The downside is what we are limited to from recommendations. If we only watch what is recommended to us then we don't see much of the other content that we may have an interest in, but YouTube doesn't know about it. An example would be anomalies; things we don't usually watch but occasionally search for.

> In theory this system should allow a person to seamlessly have enjoyable content delivered right to them. However, there is a problem with the relevance of these recommendations. First, Netflix asks you to rate content that you have viewed, assuming someone does rate every piece of content than I suppose the machine would be able to learn more about what an individual likes and be more accurate with these content recommendations. However, many people do not rate this content begging the question, how does Netflix know if you enjoyed what you have seen. I could watch ten awful pieces on Netflix and the site would just recommend other content I would dislike. Second, no matter how much the system tries to predict a person mood it cannot. Machine intelligence does not have emotions and simply cannot understand them. Maybe, I watched a romantic comedy because I was with a girlfriend and felt happy, and maybe I watched a break up movie because I was

depressed because I had just had a relationship end. Those situations that evoke two emotions that are on the opposite ends of the spectrum are based of human situations that are semi uncommon in nature. Thus, the computers generations are probably going to have a somewhat skewed interpretation of an individual's interests or desires.

For me these recommendations do not fit my interests too much because I am not consistent with my selections. I do not us Netflix too often, and if I do my selections do not make too much sense. I will jump from watching movies like lord of the rings, to movies like happy gilmore [*sic*]. Also since i do not watch tv that often by myself, when i do watch it i usually have other people with me, and their opinion gets taken into consideration on my account, which would be inaccurate because its not me who is choosing the movie.

Importantly, the reflections on the negatives of narrowcasting and Netflix were notably longer than the entries that framed the practice as a positive. This will be further analyzed in the discussion section.

Next, there were journals that viewed the narrowcasting as a part of larger trends, such as privacy, artificial intelligence, and computer learning technologies. Netflix was then critiqued as having potentially negative ramifications on overall society, not just the user. One 19-year-old participant noted that

while this computer learning may provide some convenient and useful elements to any application that utilizes such programming, there still remains the matter of privacy and identity. While computers analyzing human thought and behavior could lead to positive ramifications for education and ease of access, there still exists the common conception of computers learning too much and invading matters of privacy and security. While this may not be a matter of significance in regards to Netflix, other applications concerning more personal files and information bring these issues to light. Sensitive information becoming automated—while many companies reinforce their security—is still at risk of being compromised in some way.

Here, it is clear that the millennial participants view Netflix's practices as tied to larger social and media industry trends. As Netflix continues to recommend personalized content suggestions, the users describe their growing awareness of their data being used for overall negative purposes (despite the increasingly specialized content).

There are also larger connections to the future of society and how narrowcasting could end up hurting the population over an extended period of time. Again, these reflections offer insights into both millennial participants and the perception of the future of technology.

This artificial intelligence provided, I believe, will end up hurting our society but not before it helps our society first. When this AI comes to fruition where it is everywhere and anywhere, this will create more ease in our everyday lives basically because it is like having your brain in a computer, allowing for your thoughts to be captured perfectly. The problem that could happen with this AI technology is the lack of any need for human interaction or need for self-thinking. With the lack of self-thinking, the AI might then be able to overtake the whole of thinking for the user instead of the user generating their own content.

I am usually an optimist, but I'll say that computer learning is bad for society. From just observing how YouTube operates, advertisers seem to benefit the most from computer learning. Advertising bothers me especially when they can target my interests. It's bad enough that old media platforms like the newspaper, television, and radio are flooded with advertisements, but now advertisers can possibly use computer learning as a tool to use through new media platforms.

These quotes demonstrate that many of the millennial participants viewed the practices of Netflix as related to larger social issues such as artificial intelligence and machine learning. Both Netflix and these practices are criticized for their impact on both the individual user and society, despite their success at providing personalized recommendations. This juxtaposition will be explored in the coming sections.

Finding Three: Future Predictions

The majority of participants thought that Netflix's practices were just the tip of the iceberg of the future potential of narrowcasting, algorithms, and user-based data mining. These practices are linked together through the lens of the media industry and the users' Netflix accounts.

I think computer learning will continue in the future with the growth of artificial intelligence in more places.

I don't see how this learning is either good or bad. I think this is an on going case study for what people like since moods vary and peoples [*sic*] decision making for the next video varies. In terms of time, yes this can be time consuming since watching one video could turn into watching five videos or even twenty videos. Some benefits are that your searching is already done for you and you can find something within that subject that you didn't know existed under a different title.

In the future, I believe that computer learning will move to be more interactive and start to become even more closer aligned with the ideas of artificial intelligence.

Specifically, these reflections above offered insight that narrowcasting and Netflix's practices were the beginnings of a new standard of production in mass media. However, other participants were uncertain if these trends would eventually be positive or negative.

I believe that computer learning will only continue to grow and evolve for better or for worse. Maybe at some point the government will get involved and limit how computer learning is being used and developed. I also see any type of evolution of this computer learning to actually take some time and possibly longer than my own life span.

Computers will be able to predict and learn things that were never possible before.

It is going to change the way content is created and presented to the user, and hopefully will be put to good use to impact online video streaming in a positive way.

Except no matter what it's being used for, it will always be a double-edged sword. Safety could turn into an invasion of privacy and exciting recommendations can turn into the annoyance of being more figured out than you thought. But I do think machine learning will continue to get more detailed and individualized in the future. I also think it will just link together and use information from all of the devices a person uses, more so than it does already.

Netflix was also compared to other companies; illustrating how these same practices may be used or are already in use by other media organizations.

I think computer learning will really advance in the future, like … how Amazon is now delivering with drones in California, if it is a success it probably become worldwide in the matter of time. Computers will become faster and "smarter." I don't know how many times I've ended on YouTube longer than I originally intended because I just kept clicking more of the recommended, similar videos in the sidebar. I guess there is a chance that this technology could go too far and learn too much about our personal life but right now I can't really think of any technology that could be that extreme.

Overall, I think the learning technology of computers and things of the like will only get more accurate, and implemented into more platforms. It helps websites and advertisers target the consumer, especially since as a society, we constantly find ways to avoid advertising. We install Adblock on our browsers, DVR shows so we can fast forward through television ads, or get Netflix so we don't have to spend $20 at the movies plus sit through all of those previews. Advertisers have to capture our attention because otherwise we will find a way around it, so this type of learning helps them find out exactly what we want to see so maybe we will sit through it and hear their message.

The expansion of the scope of the journals is particularly interesting, as participants were only directly asked about their interactions with Netflix. Their inclusion of other businesses and industries suggests that the millennial participants viewed narrowcasting as a larger social and institutional trend, rather than something unique to Netflix.

Finally, many participants referenced other events, films, and pop culture in an effort to describe their hesitations about how these trends might evolve over time:

I really do think computer learning will continue in the future for a long time, even when we move to different platforms. They will become so advanced with AI, but will never have agency, or the human aspect of learning. It makes me think of the *Terminator* movies, where Skynet becomes self-aware and turns into actual intelligence. At that point we start playing God and question what actual intelligence is, with our without the human aspect. Even robots people are making are moving in that direction. They don't have to be full-size, human replicas. I saw on the Science Channel a couple days ago about body suits that help people with disabilities regain their mobility. Those suits move based on sensors in the body. I know what was a bit of a tangent, but it's still similar to computer learning.

It is the culmination of these quotes and journal entries that present a complicated look at how millennials view the future of Netflix, the larger media industry, and narrowcasting. Most participants recognized Netflix's practices and methods of data collection and linked them to larger social trends such as the reduction of privacy, growth of computer/machine learning, and even artificial intelligence. It is these connections that are perhaps most relevant to a reflection on Netflix, because of the often negative tone and criticism often levied by participants at these growing industry trends.

Reflections

The millennial participants in this study were highly aware of the technological platform and techniques employed by Netflix to collect user data and provide recommendations. The insight provided in these journals suggests that not only are participants aware of the techniques, they see potential positive and negative consequences in both the short- and long-term. While the previous section detailed the discourses found within the journals, this section will elaborate on the meaning of these findings and provide contextual information to for future research.

Despite the open-ended prompt, millennial participants relayed a nearly uniform means of reflecting on Netflix and its narrowcasting techniques. As mentioned previously, many participants identified and described the techniques and technical process used to collect information about users and provide recommendations. Most participants not only identified the process, but also were able to use the language of narrowcasting and big data analysis to do so. Terms such as "computer-learning," "artificial intelligence," and "data mining" suggest that participants are well versed in these techniques. While the term narrowcasting was never used in the journal entries, participants correctly identified the process and techniques behind it. Even those who were not exactly sure of the process of Netflix and its narrowcasted recommendations understood that they were achieved by processes of tracking and monitoring. This is critical because previous research has suggested that millennials are unaware of the processes used to collect digital data and the role it has in their everyday lives. Contrary, this study suggests that not only are millennial participants aware of these procedures, they can also identify how it influences their user experiences.

This may also give insight into why participants frequently noted that they favored the Netflix platform over other online streaming competitors such as Hulu. Netflix's personalization process, despite its lack of public information about user data, was said to be more understandable than other media platforms. This is particularly clear in the journals' connections to other media platforms, companies, and even film analogies. While there is still uncertainty associated with Netflix, there is also some confidence with the description of its process. This perhaps explains the popularity of the platform among millennials, although much more research will need to be conducted to support this hypothesis.

Findings on the user experience and future predictions provide us with examples of how millennials view the role of the narrowcasting process moving forward. While most participants recognized there were some positives, including saving time, having a more meaningful media experience, and tailoring digital media options to personal preferences, more participants argued that the Netflix platform represented a long-term negative turn for the role of technology in everyday lives. Again, unlike previous literature that suggests millennials were born into a digital environment and thus are less critical of digital platforms and consequences, these participants demonstrated that they were both aware and concerned regarding the impact that these technologies may have on the future.

Furthermore, although millennials are identified as being concerned with the implications technology may have on their own lives and not on society in general, these participants nearly all reflected on how Netflix's practices are indicative of larger trends—and other companies—and may potentially negatively impact other people. While more quantitative work would need to be done to generalize these findings to the larger millennial population, this does suggest that effects are viewed as both individual and social within the Netflix platform.

Because most participants identified the processes used by Netflix to create recommendations, they could also then critique the smaller technological elements that made up the larger effects. For example, four participants identified that many of the techniques used on Netflix are also the foundations of artificial intelligence technologies. Therefore, the journal was utilized as a means for looking at larger trends in digital data and technology. As with previous research, millennial participants in this study failed to take a hard stance on whether these technologies would have a positive or negative overall effect on either individual users or larger society. While not generalizable to the entire millennial population, these findings provide insight into the mindset of the generation. For example, 20 participants identified how narrowcasting may have both pros and cons (the remaining seven did not discuss effects). However, this may be more reflective of the journal prompt than an actual finding regarding the millennial generation. The prompt asked participants to reflect on their experience after looking over their Netflix account.

As a result, this implies that they would find both positive and negative aspects to the platform.

Recruitment is also one possible reason for these findings. University Institutional Review Board policies on this project only allow for limited data collection using journal analysis. This means the analysis cannot make larger conclusions regarding gender, location, major, language, citizenship, or time with Netflix account. These are all considerations that need to be placed in the future. Other future research should identify how other age groups interact with Netflix and narrowcasting on other media platforms. Many participants here argued that sites such as Amazon, Hulu, and Facebook similarly integrate user content into recommendations. However, other methods such as focus groups or in-depth interviews should also be considered to clarify some of the original themes found within this study.

The role of narrowcasting in creating programing and platforms for targeted audiences is a growing trend in the digital media industry. As a result, it is critical to understand how various audiences react to these evolving norms. The millennial generation, often viewed as narcissistic and lacking critical skills, not only recognized narrowcasting as a part of the Netflix platform but also had strong opinions regarding its long-term consequences in society. While not a generalizable finding, this seriously challenges previous estimates and theories regarding the influence of narrowcasting on young people and the digital media environment.

Journal Prompt

Thank you for participating in this journal study on Netflix. Below you will see a journal prompt. Please type your journal entry in the space below. When you have finished your submission, please click "complete" at the bottom of the screen. This will submit your entry and conclude your participation in the study.

Before starting, login to your Netflix account and look at the recommendations for genres, titles, or actors. Consider how well they fit your interests, viewing history, or preferences. In the space below consider and answer the questions in narrative form. Use any examples, experiences, or other content to support your answer.

How well do Netflix's recommendations fit you? How do you think Netflix makes these recommendations or knows about viewing interests? Are these recommendations a positive or negative aspect of the site?

NOTES

1. "Netflix Prize," *Netflix*, n.d., accessed October 1, 2014, http://www.netflixprize.com/rules.

` 2. Tom Vanderbilt, "The Science Behind the Netflix Algorithms That Decide What You'll Watch Next," *Wired*, August 7, 2013, accessed September 1, 2014, http://www.wired.com/2013/08/qq_netflix-algorithm/.

3. Suchan Chae and Daniel Flores, "Broadcasting versus Narrowcasting," *Information, Economics, and Policy* 10 (1998): 45.

4. Vanderbilt, "The Science Behind the Netflix Algorithms."

5. *Ibid.*

6. *Ibid.*

7. Blake Hallinan and Ted Striphas, "Recommended for You: The Netflix Prize and the Production of Algorithmic Culture," *New Media & Society* 18.1 (2014): 117–137.

8. Tony Cheng-Kui Huang, Ing-Long Wu, Chih-Chung Chou, "Investigating Use Continuance of Data Mining Tools," *International Journal of Information Management* 33.5 (2013): 791–801.

9. Steve Lohr, "Netflix Cancels Contest After Concerns Are Raised About Privacy," *New York Times*, March 12, 2010, accessed October 1, 2014, http://www.nytimes.com/2010/03/13/technology/13netflix.html.

10. Dustin D. Berger, "Balancing Consumer Privacy with Behavioral Targeting," *Santa Clara Computer & High Technology Law Journal* 27.1 (2010): 3–21.

11. Felix Salmon, "Netflix's Dumbed-Down Algorithms," *Reuters*, January 3, 2014, accessed June 10, 2014, http://blogs.reuters.com/felix-salmon/2014/01/03/netflixs-dumbed-down-algorithms/.

12. Megan Mullen, "The Rise and Fall of Cable Narrowcasting," *Convergence: The International Journal of Research into New Media Technologies* 8.2 (2002): 65.

13. *Ibid.*

14. Beretta E. Smith-Shomade, "Narrowcasting in the New World Information Order: A Space for the Audience?" *Television and New Media* 5 (2004): 75.

15. *Ibid.*

16. Susan Tyler Eastman, Sydney W. Head, and Lewis Klein. *Broadcast/Cable Programming: Strategies and Practices* (Belmont, CA: Wadsworth, 1985).

17. *Ibid.*

18. Smith-Shomade, "Narrowcasting in the New World Information Order," 75.

19. Eileen Meehan, "Why We Don't Count: The Commodity Audience," in *Logics of Television: Essays in Cultural Criticism*, ed. Patricia Mellencamp (Bloomington: Indiana University Press, 1990), 117–37.

20. Smith-Shomade, "Narrowcasting in the New World Information Order," 75.

21. Syed M. Khatih, "The Exclusionary Mass Media: 'Narrowcasting' Keeps Cultures Apart," *Black Issues in Higher Education* 13.11 (1996): 26.

22. Peter Ludes, *Convergence and Fragmentation: Media Technology and the Information Society* (New York: Intellect Books, 2008).

23. R. Kelly Garrett and Paul Resnick, "Resisting Political Fragmentation on the Internet," *Daedalus* 4 (2011): 108.

24. *Ibid.*

25. Zvezdan Vukanovic, "Global Paradigm Shift: Strategic Management of New and Digital Media in New and Digital Economics," *International Journal on Media Management* 11 (2009): 200.

26. Philip N. Howard, "Deep Democracy, Thin Citizenship: The Impact of Digital Media in Political Campaign Strategy," *The Annals of the American Academy of Political and Social Science* 597 (2005): 153.

27. Zvezdan Vukanovic, "Global Paradigm Shift," 81.

28. *Ibid.*

29. Helen Wood, "Television Is Happening: Methodological Considerations for Capturing Digital Television Reception," *European Journal of Cultural Studies* 10.4 (2007): 485.

30. "Narrowcasting Revenues Expected to Triple by 2009," *TechWeb*, 2009, accessed June 10, 2014, https://business.highbeam.com/138350/article-1G1–127480743/narrowcasting-revenues-expected-triple-2009-revenues.

31. "Research and Markets Offers Report on '2007 Trends to Watch: Media & Broad-

casting Technology,'" *Marketsensus*, 2007, accessed June 10 2014, http://marketsensus.com/2007-trends-to-watch-media-broadcasting-technology.

32. Jennifer Gillan, *Television and New Media: Must-Click TV* (New York: Routledge, 2010).

33. "Advertising and Marketing Companies: How Audience Size Affects Word of Mouth," *Marketing Weekly News*, April 2010, accessed June 10, 2014, http://www.mckinsey.com/insights/marketing_sales/a_new_way_to_measure_word-of-mouth_marketing.

34. D.T.Z. Mindich, *Tuned Out: Why Americans Under 40 Don't Follow the News* (New York: Oxford University Press, 2005).

35. Debora S. Vidali, "Millennial Encounters with Mainstream Television News: Excess, Void, and Points of Engagement," *Linguistic Anthropology* 10 (2010): 275.

36. Alison N. Novak, "Millennials, Citizenship, and *How I Met Your Mother*," in *Parasocial Politics: Audience, Pop Culture, and Politics*, ed. Jason Zenor (New York: Lexington Books, 2014), 200.

37. Nick Couldry, Sonia Livingstone, and Tim Markham, *Media Consumption and Public Engagement: Beyond the Presumption of Attention* (London: Palgrave MacMillan, 2007).

38. Laura Harvey, "Intimate Reflections: Private Diaries in Qualitative Research," *Qualitative Research* 11 (2011): 664–684.

39. Niall Bolger, Angelina Davis, and Eshkol Rafaeli, "Diary Methods: Capturing Life as It Is Lived," *Annual Review of Psychology* 54.1 (2003): 579–619.

From Interactive Digital Television to Internet "Instant" Television

Netflix, Shifts in Power and Emerging Audience Practices from an Evolutionary Perspective

Vivi Theodoropoulou

This essay examines users' experiences with Netflix, focusing on the emerging viewing habits, consumption patterns, and preferences of early adopters in the United Kingdom, and seeks to identify what these users appreciate and dislike about streaming video technology. More pointedly, this essay relates Netflix early adopters to initial users of Sky Digital, a pioneering interactive satellite digital television (DTV) service of the early 2000s and a Netflix precursor of sorts. Promoted with the slogan "Watch what you want, when you want"—one almost identical to Netflix's messaging today—Sky Digital quickly grew into the most popular UK DTV service before being overtaken by Freeview, the terrestrial DTV operator, in 2007.[1] Through a comparison of Netflix and Sky Digital, the essay illustrates how television and its audience are changing, how users respond to digital content, and the ways in which new habits develop around technological features offered by popular content services.

The essay primarily relies on qualitative research consisting of 12 interviews with UK Netflix users conducted between late 2014 and 2015. Linking broadcasting with online television, I juxtapose these interviews with older findings from quantitative and qualitative research on early Sky Digital subscribers conducted in the early 2000s. Netflix and Sky Digital were released

182

to the public at different points in time and present different technological platforms and distinct ranges of content. Yet, the two platforms still share important qualities—including increased consumer choice and customized, targeted content offerings—and, notwithstanding modern audience practices, produce similar perceptions and habits among users. As such, this essay demonstrates how certain consumption patterns and preferences persevere over time, and that shifts in audience behavior always co-exist with continuity. Change does not happen swiftly; rather, it is iterative and progressive in nature. In other words, technological knowledge, invention and innovation, and the use of new technologies all develop in "path dependent" ways.[2] While they are rooted in the embrace of new practices and in our changing social and life circumstances, they are also based in past practices and experience.

Netflix and Sky Digital's Affordances and Adoption in the United Kingdom

Netflix launched in the United Kingdom in January 2012. According to research from Ofcom, one in ten households had a subscription by the end of 2014.[3] This signals that Netflix is at the early adoption stage of Everett Rogers's diffusion of innovations model.[4] Research shows that broadband television use in the United Kingdom continues to increase significantly, and on-demand services such as BBC iPlayer and Netflix are at the forefront of such a development.[5] Beginning in the United States in 1998, initially as a DVD-by-mail rental service and then online video store providing access to films and television series at a relatively low fee, Netflix has shifted into a successful digital distribution platform that produces and streams successful content from the past and the present.[6] On Netflix, subscribers can access content when and how they want, thus turning the standard continuous television schedule, or broadcasting "flow," into a thing of the past.[7] Despite not publicly releasing specific ratings data, Netflix is now also famous for the quality and success of its in-house productions, largely credited to the platform's algorithm, and manifested in its Primetime Emmy, Screen Actors Guild (SAG), and Golden Globe nominations.[8]

In academia, policy cycles, and the industry, Netflix and other post-network services are thought to be significantly altering consumption patterns and audience habits and changing the face of home entertainment.[9] For instance, Amanda D. Lotz explores the potential end of television viewing as a shared experience, while William Uricchio and Jenna Bennett suggest that modern television lacks the ability to shape daily life on a mass scale because of audience segmentation and diversification.[10] Likewise, Elizabeth Evans details the ways in which online television consumption—including increased time- and place-shifting—may change our perceptions of what "television" actually is.[11] Indeed, such explorations are needed to assess the change that new online television formats might bring—not just on production and distribution, but also for audiences and society at large. Netflix is at the forefront of such change due to its exceptional delivery system, the targeted, uninterrupted, and on-demand viewing it allows, and the overall control and flexibility it gives users.

Yet, many of the changes proclaimed by Netflix were also proclaimed by DTV in the United Kingdom with the launch of the first DTV service, Sky Digital. Introduced at the turn of the century, Sky Digital was a subscription-based multichannel broadcasting service offering a multitude of entertainment options, including single-genre thematic channels as well as what were then considered groundbreaking, Internet-like interactive facilities (such as the "Open...." platform offering access to electronic games, emails, banking, and shopping).[12] Sky Digital's unprecedented choice of channels and content gave the user the freedom to focus on her/his favorite type of content or genre in one of the many thematic channels (including film, sports, news, documentary, kids, comedy, music, and much more). Sky Digital also offered a relatively limited option for users to intervene in the broadcasting process and manipulate the television schedule via time-shifting and (near) video on-demand features such as Pay-Per-View (PPV), TiVo, and Sky Plus.[13] Though such services nowadays might be considered limited and primitive technologically, at that time they were seen as innovative and cutting edge. In a sense, then, DTV of that time could be considered an unsophisticated predecessor of Netflix.

Digital Television, Its Early Audience and Some Parallels with Netflix

When DTV was introduced in the United Kingdom, I studied how the first generation DTV audience responded to this new technology through a nationwide survey research of Sky Digital subscribers, followed by 15 qualitative in-depth interviews with DTV users.[14] My quantitative research showed that early users switched over from their analog television to DTV because of the wider choice of channels it offered (77 percent of users), the improved picture and sound (68 percent), the availability of more sports (49 percent) and, interestingly, more film channels and film-related programs (32 percent).[15] In this respect, early DTV users exemplified an early preference towards thematic, singe genre channels, indicating an increased turn toward segmentation but crucially also toward customization and personalization.

The timid acceptance of the PPV among DTV users also indicated this trend.[16] Appreciation for customized and personalized content noticeably manifested during the early days of DTV and has clearly continued to evolve, and ultimately, has been further fostered by algorithmically driven platforms like Netflix.

My research at the time also showed that for selecting and subscribing to the service of Sky Digital in particular, special offers, the free digi-box giveaways, and more competitive pricing were important incentives for users. It is also important to note that among the vast array of programs available to them, films were the most popular type of content early DTV users liked to watch (just over 69 percent of participants), and this was closely followed by comedies (just under 69 percent), sports (62 percent), and dramas (51 percent).[17] This audience preference towards films and series is something that persists through the years and something that Netflix is capitalizing on today. On the whole, increased channel choice, a wider range of content, the introduction of more thematic channels, and affordability were the key attributes driving the take-up of Sky Digital in those early years.[18]

Similar attributes such as low cost and perceived high value for the dollar, increased choice, quality content, and thematic programs (films and series) are today the forces behind Netflix's appeal. In addition, the assumed "user-friendliness" of the service is a significant positive attribute for Netflix, as most interviewees attest. Signs of audience emancipation, and customization of viewing, together with the avid preference towards films and series, were all evident since the early days of DTV—as they are evident, to an even larger degree, among Netflix users today. In what follows, I discuss how the appreciation of choice, control, content, and low cost matters for Netflix users and provide examples of their experiences. First, I provide a short description of my Netflix research methodology and sample, and discuss my Netflix interviewees' key demographic characteristics, their views on Netflix's cost and content, and their social and technological context of viewing.

Netflix Interviewee Profile

Netflix's diffusion in the United Kingdom demonstrates relative fast adoption rates and, in this sense, a similar path to the very quick adoption of DTV when it was introduced in the country in late 1998.[19] This fast adoption of Netflix indicates a certain attraction of subscribers to the service, and this essay identifies some of the reasons for it. Particularly, I discuss the appeal of Netflix, how UK users in their 30s and 40s consume content on this digital streaming platform, and why they subscribe to the service in the first place. The study draws on qualitative semi-structured and structured face-to-face and online interviews conducted between end of 2014 and 2015. Given that Netflix does not readily provide public information about its subscribers and the fact that the service is admittedly at an early (if swift) stage of adoption in the United Kingdom, early adopters were not particularly easy to track. However, 12 users, recruited via acquaintances and snowball sampling, were interviewed in total.

All subscribers of Netflix interviewed are in their 30s or 40s and in full-time work.[20] They are of a high educational level (college degree and above); seven are men and five are women; one is single, three live with their partners, eight live with their partners and young children. Eight of the interviewees constitute four couples, though they were not interviewed as couples but as individuals. These interviews illustrate that most participants subscribed to Netflix because of the content offerings and the affordable cost. Regarding the quality, most interviewees find Netflix to offer a very high standard of content. Most interviewees also suggest that the Netflix free trial initially inspired them to try the service and that they are generally satisfied with the price of a monthly subscription. Thus, respondents say that Netflix offers strong value for the cost and highlight this as one of its primary assets. Much like in the case of Sky Digital and early DTV offerings, we see can see here that cost and free offers are crucial in "catching the consumer."[21]

Elizabeth Evans and Paul McDonald suggest factors such as range of content, cost, and ease of use as most essential in shaping consumers' attitudes toward digital distribution services.[22] Moreover, Phillip Napoli refers to "the easily searchable cornucopia of Netflix" and its great navigability as factors that enhance its use.[23] In line with such findings, most interviewees also note that Netflix is user friendly and easy to use, with a clean interface. Having said that, most participants do not find the recommendations or personalization system very helpful but rather just okay. Yet, many remained fascinated by the system; they are surprised at how frequently its suggestions fit their personal tastes. In terms of the interface used for viewing, just over half the participants watch Netflix on their television mostly and on their laptop, and the rest watch on their laptops and/or tablets but not on television. One couple also uses their smart television for Netflix viewing. As far as the social context of viewing is concerned, most interviewees watch Netflix alone or with their partners, family, and/or friends, while only two watch only by themselves. Having briefly described their basic demographics and family status, and quantified/ summarized the interviewees' levels of satisfaction with the content and choice offered, the cost, and user-friendliness of the service as well as the technology they access Netflix from, I proceed with discussing how they use it, what they like about it, and how they develop habits around it, in more qualitative terms.

Instant Television: On-Demand and Personalized Viewing

Aligning with Hye Jin Lee and Mark Andrejevic's analysis of second-screen applications and contemporary media devices, Netflix's status as a time- and place-shifting technology allows viewers to watch the programs they want, on their terms, avoiding the constraints of standard television schedule while they can also skip the standard television advertisements.[24] All participants are attracted to Netflix principally because of its on-demand characteristics. They are content with the fact that they are liberated from the restrictions of the traditional television schedule. This characteristic of Netflix is taken up by users so as to fit their lifestyle and everyday life circumstances, and this is a reason why Netflix is so successful with subscribers. With this obvious flexibility, the service does not disrupt such everyday patterns or structures. In other words, Netflix permits for a firm and easy domestication.[25] Participants' age and working circumstances, their family life, domestic activities, and free time all play a role in how they use Netflix. Likewise, Netflix's technological features, control over the scheduling, and the flexible viewing it allows play a role in enabling its easy incorporation in daily routines. As Leo, a 40-year-old busy professional working long hours, commented, "You watch whenever you want and wherever, at home, in the garden.... It's convenient to have control. Who's watching standard TV these days? Only my mother-in-law. She's got the time. I don't. I don't have the time when the TV wants. I have the time when I have the time."

Leo believes that a service like Netflix is tailored for people his age, generation, and busy lifestyle. He went on to justify how, for people who work long hours in our days, Netflix provides an emancipatory potential and helps them control and structure their busy lives better. He said:

> You have to be over 35–40 [years old] to watch Netflix. Because these are the people who always want to watch TV but never can cause the standard TV program does not fit their daily life structure. But now they can watch when they want... .With standard TV you might want to watch but often you cannot. I can't watch when I'm back from work because I have to walk the dog. But I can when I'm back from walking the dog. I'm more relaxed about it now with Netflix, because I can.

Here, Leo gestures toward how Netflix caters to certain needs created by the new paradigms of work and labor and contemporary lifestyles that leave little freedom, in the sense of time, to engage with activities one likes. These also leave little room for unplanned or random activities, as Neil, another interviewee, mentions. Neil added to the discussion on this other dimension of Netflix viewing: that of planned, quality viewing and selectivity. Neil does not have a standard television set. Instead, he said that having instant access to television content means he can be more focused and selective with his viewing and be freed from the standard schedule at the same time:

> Having instant access to TV content, it means choosing to see [a program], when I want to see it. I do not like the idea of watching just whatever is on, or to have to live by a TV schedule, as is normal with linear delivery.... I do not use standard television, and have not used it for a long time. I would say that I am more likely to watch things I'm not really interested in on standard TV, which I dislike.

This is similar to the comments of Lenny, a father of two, who noted that—even as he enjoys bingeing—he now plans and pre-selects his viewing on Netflix, meaning that his viewing of random programs on standard television has decreased. Mimi, a new mother, also enjoys the control over viewing that Netflix allows; yet, she identifies things she now misses out by not watching normal television. Her circumstances, much like those of Leo, mean that on-demand viewing is preferable due to its convenience. In her newly formed family structure with a new baby and new responsibilities as a mother, Mimi suggested that Netflix enables structured viewing and planned time to relax with television. She assured that being able to watch something when she can is "important because we have a set routine at home and on normal TV there's nothing on. Obviously the drawback is that I don't watch normal TV and am behind on news."

The participant's last remark is indicative of a change in viewing patterns taking place when viewers focus on Netflix content alone and give up—or significantly decrease—their traditional viewing. Being caught up in the flow of the service and thus in touch only with the genres available through it, viewers miss out on other types of content on standard television, including news and current affairs. Kim in particular strongly regretted not keeping up with current affairs and news programs as she used to before her subscription to Netflix. She explained:

> I'm less and less aware these days of when things are actually on [standard] TV. It's like the Netflix style of viewing seeps into all viewing.... [A]nd I find I am missing things that are good that are actually being broadcast, as I don't check anymore.... Now it's less likely I'll come across something I wasn't planning to watch, and also if I am honest, I watch less news and current events ... the more I think about it, having control over the viewing with Netflix, I don't think it is a great thing and the more I think about it scheduled TV has functions we are now missing, most notably news and current events.

Other interviewees also stressed this fact, noting how they do indeed miss some of their old viewing habits and practices set around programs they watched on standard television. Some complained about missing sports, like Mike, who said, "It has the greatest series ever. It's a shame it does not screen any sports. They should get sports in there." Meanwhile, Eve, while talking about herself and her husband, echoed Mike's sentiments to a degree, "On

Netflix you don't have the news, that I now watch or follow online, and my husband misses his sports, I guess." On the issue of sports and news, Chris, a big fan of Netflix, made a clear delineation between the platform and the rest of television; he stated that "normal television for me is about sports and about news. Other than that I don't think I watch normal television. It doesn't particularly bother me that these are not on Netflix. I also use a lot of online forums to keep up to date on sport. But virtually anything that's not news or sport I would watch on Netflix." We see here how Netflix is beginning to displace traditional broadcasting and "normal" television. For some users like Chris, traditional television remains a part of their lives only through the content that Netflix does not offer like sports and news. Also, it seems that such users are very content with the range of content available on Netflix and equate it to normal episodes and series, turning "normal" television into something different, or an alternative.

On the issue of control, interviewees like Kim realized that they have now excluded certain things from their daily lives and television habits, largely because of Netflix. Likewise, Gareth expressed ambivalence about the value of control over his viewing. He related this outlook to his old viewing practices and his age/generation:

> Maybe, it's a good thing for us to have less control over what we view every once in a while. I cannot imagine this is a popular viewpoint, and it has probably got something to do with my age, but there's also a wider political point to be made here about the extent to which we're losing control over all aspects of our lives, except as consumers.... If you only grew up with three channels on the TV and had to wait for a week for the next episode of your favorite show then TV on-demand is nice but you don't think of it as a human right!

Eve also agreed with Kim and Gareth that having less control over their consumption is not necessarily bad. Still, Eve noted that she is often irritated by the fact that she and her husband cannot agree on what to watch on Netflix, and that she misses certain elements of her old viewing habits. She contemplated how much better it used to be in that sense:

> As I said, with standard TV we used to watch whatever was on in the evening when my partner would come home. This I liked because we'd spend time together. I'm used to watching TV to relax and like easy to digest stuff. So I was not planning my viewing. I just like hanging out there on the couch together with my partner. Now, yes indeed, I'm more selective, feel more in command of my viewing. But still I miss that being together as a couple in front of the TV and watching whatever with Mike. It wasn't like that with Netflix at starts. It's since he got so hooked on *House of Cards* [2013–], that I do not like, that we separated in front of the TV. So now we watch on our own on our laptops. I don't watch that anyway.... You are in more control with Netflix and I'm more independent as I said. But the standard TV schedule was not a bad thing either. As I said, I really did enjoy watching whatever was on, and not having to always do the choices. That's not really bad. I sometime miss it.

It again should be emphasized that Netflix users celebrate their enhanced freedom and control over the schedule, and their personalized viewing allowed by the platform's on-demand capacity. In fact, this freedom and control, without technological or time constraints, is what distinguishes Netflix from DTV and other services of the past. Netflix and other online streaming platforms catalyze a shit away from the broadcasting schedule or what Uricchio calls "the programming-based notion of flow ... to a viewer-centered model."[26]

However, as some of the above comments illustrate, interviewees sometimes miss this flow provided by broadcast television. They appreciate the planned viewing of Netflix's algorithm, but some reminisce about the unplanned and unstructured viewing facilitated by traditional television. Thus, while interviewees form new Netflix-related habits, they still retain some of their old habits, or even develop new ones outside of Netflix (such as following news and current events online). These consumption habits align with Yu-Kei Tse's claim that despite the increasing success and "growth of online platforms and personal media devices ... it is inappropriate to assume that the media environment and audiences just switch from one end (broadcasting-mass audiences) to another (narrowcasting-individualized)."[27] Or, as I would like to suggest, by linking Netflix users of today with interactive DTV users of the early millennium, change in viewing practices and habits coexists with continuity. User habits—both within the same and across different media platforms and distribution modes—tend to shift in a gradual, evolutionary way.[28] This gradual shift of practices appears "as a result of a constant interaction between the user and the technology."[29] Lisa Gitelman proposes that a new medium and all its "supporting protocols" (that is, the technological standards but also the norms about how and where we use it) develop slowly, and through time they "become self-evident as the result of social processes, including the habits associated with other, related media."[30] Indeed, as Philip Palmgreen and Jay D. Rayburn assert, consumption patterns and the gratification users expect and gain from media usage are based largely on past experience—or as Terje Rasmussen similarly claims, "old new media practice structures new media practice."[31]

Content and Viewing Experience

In terms of content, interviewees are very satisfied with the quality of films and series available on Netflix. Even if a few complained about the limited range of accessible material, they all seem happy with the high standards of the content and, as Maya described, the "almost cinematic experience" the platform allows. Many of the participants mentioned the numerous serials available on Netflix (such as *Breaking Bad* [2008–2013], *House of Cards*, *Weeds* [2005–2012], and *Orange Is the New Black* [2013–]) as key selling points of the service. For instance, when commenting on the quality content, Maya noted that "*Breaking Bad* was the reason we got on Netflix too. Great cult series that shows that television is neither dead, nor lacks quality and imagination in production. Sometimes."

Regarding the viewing experience, two points of interest arose in my research findings: (1) the contradictory feelings of empowerment over controlling the schedule and guilt over extended binge-watching; and (2) the association of on-demand viewing—on one's own schedule and with one's preferred content—with relaxation and entertainment.

Importantly, both of these are also characteristics of the early DTV era of enhanced choice. On one hand, respondents familiar with older television eras also tended to be a bit defensive when watching too much and concerned with the quality of television. On the other, they generally associated viewing with leisure, relaxation, and entertainment, considering it a "laid back," low-demand, easy activity in which to engage. A 30-year-old male DTV subscriber, for example, explained why he rejected the interactive services on Sky Digital: "It's too much hassle; you really want to sit back and just watch whatever's on TV." Similarly, a female 35-year-old DTV interviewee and mother of two whose parental responsibilities defined her media consumption noted that she did not use the enhanced interactive services because her exhaustion meant she preferred to just watch programs. She stated that she used television to relax, unwind, and to be entertained without having to make an effort: "I'm watching a program, I don't want to read about it!... I just can't be bothered.... I haven't got the time or energy.... I'm usually watching the telly half switched off. So to actually add any more information to what's coming in would just be too much for me." Interestingly, users of the post-network television era also link their (online) viewing with needs and expectations they previously associated with television, as will be shown in the next section. They still consider their online viewing a leisure activity that takes place when they have the time, in order to relax and escape from the concerns of the day, without putting too much effort to this practice.

Binge-Watching and "Guilty Pleasures"

On-demand bingeing is a novelty that signals a change in viewer habits that scholars are beginning to acknowledge and examine.[32] At the same time, however, one could say that watching, for example, a week's worth of popular British soap opera *Eastenders Omnibus* (1994–2015) episodes when they screen, in sequence, every Sunday on BBC One, is not much different as interviewee Eve mentioned. The fundamental logic of continuous viewing is the same between the two examples, but there is still a crucial difference—Netflix gives users the control of when, where, and how many episodes they binge.

Many of the contemporary interviewees binge content on Netflix. Like most viewers, they like not waiting a week for the next episode, but now watch for extended periods of time, or, as Tilly confessed, have "lots of late nights." Often, with this pleasure come feelings of guilt for viewing too much—and sometimes with embarrassment in admitting it, thus turning binge-watching into what Debra Ramsay calls a "guilty pleasure."[33] For example, even though Kim admitted that one of the things she appreciates most on Netflix is "being able to watch as many episodes as you like in a night—bingeing," when asked to say how many hours she devotes to it she responded: "I don't know. Don't like to think about it!" Maya also expressed the same kind of underlying guilt for bingeing, even if she considered her Netflix viewing as better than the "vegging out" one does on standard television.

Leo, in describing his bingeing patterns, sounds like he is trying to justify himself. Arguing for his pleasure, he stated (in a somewhat defensive tone), "I don't need to download, I just watch. I do binge a bit yes. I can watch 5–6 episodes in a row of How I Met Your Mother (2005–2014), for example. Why not? I am drawn to it. I can leave it; but also, it's a way of being lazy and doing nothing. I work a lot! I don't have any regret. I can do that. Why not?" Chris on the other hand, confidently explained that he happily binges because his viewing always occurs past 10 p.m., long after he has completed his daily tasks. Chris has fixed a routine so that he can watch unobstructed from any preoccupation in a carefree mode:

> Oh yes absolutely. I do binge. I like it. Yes time flies and you watch a lot. I've been watching things till 3 a.m. in the morning. But I like it. It doesn't matter (devoting so much time to TV). I watch it at night; I don't watch it at normal television times so it does not intervene with the tasks of the day. I watch 10 p.m. onwards. My son's in bed. Dishes are done. Tiding up is done. Time to relax.

Here, then, we see how Chris domesticates Netflix so it will not intervene or obstruct daily tasks or patterns, and will give him the chance to binge-watch, guilt-free. Binge-watching in general "recasts our relationship to televisual time," affirming our control over media content.[34] Despite associations with guilty and out of control watching, users like Leo are "drawn to" binge-watching. This mode can also improve the viewing experience, particularly in attention to the narrative of what is being watched. It reveals a kind of commitment that can be particularly appealing to fan communities.[35] In this sense, as Chuck Tryon notes when reviewing Netflix's promotional discourse, bingeing can be characterized "not as a passive activity but as one aligned with active viewing practices, as a way of managing one's time in front of the television rather than succumbing to a television schedule."[36]

Furthermore, much like Chris's above quote indicates a process of domesticating Netflix, DTV subscribers also regularly found shortcuts and ways of incorporating their viewing into the web of daily activities. At that time, new digital video recording technology TiVo allowed DTV users to pause programs while watching and record one channel at the same time they were watching another. Of course, such a novelty was met with strong enthusiasm from those who could afford it. For instance, a 34-year-old DTV interviewee and mother of four explained what a relief the technology provided her in terms of juggling her obligations as a mom and as a fan:

> Well, I'm watching television in the afternoon; when the children are around I pause things in the middle of what I'm watching and then I'll continue watching it after I've helped them with what they're doing or after I've put the vegetables on to boil or whatever.... In order to cook the children's dinner I don't have to miss what I'm watching. I love it. And it also means that rather than saying to the children "Shhh, just let me watch this bit," I just press pause and they can say what they need to say and run off and carry on with what they were doing.

It was clear at the time that the more new gadgets DTV users could afford the more freedom they had in structuring their viewing in a productive fashion. Again, such freedom has increased further for Netflix users—as Chris implied above—and at a fraction of the cost.

Viewing Ritual and Associated Needs

Aside from bingeing, most interviewees engage in another pleasurable ritual of viewing: that of watching Netflix with their partners (or by themselves) in the evening after a long day's work. This is their "relaxation time," and is when they usually watch more television series than films. Such rituals are and have always been common with each generation of broadcast television. Most of the early DTV interviewees made similar linkages between television and relaxation in the early 2000s. These interviewees to a large extent rejected the supposedly innovative and interactive services of DTV exactly because they were not deemed compatible with the relaxing, low-demand experience of television. As a 30-year-old male interviewee told me at the time, "Television is there as a relaxing tool. You get home in the evening and it's there to entertain you. In effect, you have to actually physically do something if you're interactive, rather than the relaxing element of it, which is what television basically is."[37] In another article, I reviewed findings from this same research, discussing how differently those early DTV viewers perceived television and the computer/Internet.[38] There, a 43-year-old male made a clear distinction between the two, stating, "When I want to come and sit down and relax, I'll come down and sit down and watch the television. I can't relax if I use the computer. That's the difference, I think."[39]

Today, consumers view computers, tablets, and smartphones as more than tools for working and administrative purposes. Indeed, these tools are increasingly used not only to communicate via social media and messaging platforms, but also to watch media content; in short, they are regarded as entertainment and relaxing tools as well. Nonetheless, as the above interviewee suggested, watching television—from whichever access point—is still associated with the need to be entertained, to relax, and to unwind.

Much like early DTV users, Netflix users engage in what Lotz calls "cocooning," a practice that emerged well before the advent of DTV—that is, staying at home to be entertained with their family and/or friends by watching content on the streaming service of their choice.[40] In these instances, interviewees prefer television series to films. As Neil explained, "I love movies, but most of the time I prefer series, because I watch Netflix to relax after work in the evening, and I'm looking for light entertainment." Other participants, such as Mimi and her partner Bob, confirmed this experience, noting that the shorter nature of individual episodes means that they are easier to digest, less demanding, and thus in accordance with a relaxing mood.

Maya revealed the viewing customs she and her partner Lenny have developed around Netflix: "Series watching on Netflix has become a ritual on weeknights with my partner as we do follow two to three serials and that's the time we 'block' to sit down together and relax." She further explicated the premises on which such rituals are built: the need to be social and share time with her partner, but also the need to relax and be entertained in a relatively short period of time with undemanding yet good television. As she stated, "Low demand, shared time which helps to relax after long tiring days. Serials on Netflix are also good alternatives to films as an episode is shorter than a film so can fit in an evening without feeling that you totally abandon quality, like, for example, you do when you vegetate in front of whatever is on TV."

Similarly, Kim noted that she and her partner watch Netflix for a couple of hours a night: "after everything is done this is our down time!" She explained how their viewing fits in their domestic practices and how the couple has dinner in front of the television set, watching Netflix programs after the kids have gone to bed. As she said, "We are trying to eat earlier with our kids but our habit has been to get them dinner, get them to bed, and then chill out finally in front of the TV with our dinner—not healthy! But a quick route to relaxation, and not time to have dinner and then watch, so we've been combining." Other interviewees expressed an appreciation for this practice of eating dinner or "chilling post-dinner" while watching Netflix. Yet again, this is a practice that is very familiar for most audiences, regardless of the technology or viewing platform. As such, some viewing rituals remain relatively consistent to older eras of television. For Netflix users, the social context of viewing is still important; they still prefer to watch television with partners, family members, or friends. However, exactly because of Netflix's flexibility, it can be also regarded as a personal medium—especially when disagreements over what to watch occur like in the case of Eve and Mike. Netflix still creates notable changes in consumption, with users now free from traditional scheduling. Changes have also taken place regarding the spatial context of viewing. Though largely remaining within the private space of one's home, viewing can now increasingly take place in public space (such as the office, as Leo confesses, the park, public transportation, and more).

Conclusion: Content Is Still King

Today, television is at a crossroads. Technological developments have brought significant transformations, creating new ways of watching online, on-demand, and on the go. Yet, broadcasting and traditional viewing also persist—even if they are declining—and are still favored by a notable segment of the audience. As this essay illustrates, technological change and the proliferation of online streaming platforms such as Netflix have generally allowed for more choice and control for users. This choice and control manifests in personalized, customized viewing and ultimately eliminates the constraints of the broadcasting schedule. Yet, the choice provided by Netflix also cuts off users from content not available on the platform, including news, sports, and current affairs programming. Interviewees express how these additional programs are sometimes excluded from their "new" habits altogether, but more commonly they persist on standard television (watching sports) or develop on different media altogether (e.g., reading the news online). Therefore, as expected, changes are not necessarily abrupt or revolutionary, but rather slow and gradual.

Meanwhile, the comparison of emerging user practices in the DTV and Netflix eras demonstrated that some differences but also some similarities between the two groups exist after all. Offerings like Sky Digital first purported to allow viewers to "watch what they wanted when why wanted it," creating what was then a substantive increase in content selection in comparison to the analog service. As a 41-year-old male noted at the time,

> With digital TV you get such a choice to watch, to the documentaries, to the music, to the entertainment, to films, and you can watch it when you want to. You can usually find something in those categories if you want to watch television, to pick and sit down and relax and watch. But we won't normally go to the television to watch a

particular program. We will normally go to the television and I'll flick through and see what's on, find something we want to watch and just switch to that channel.

The experience of watching Netflix is different to DTV in the sense that "watching what you want when you want it"—or, to be more accurate, watching what you want from the content available on the service—now takes a more accurate meaning; users literally acquire control over the scheduling through on-demand, personalized functions. It is even more emancipatory, giving users freedom in not only what to watch but in *how much* of it they want to watch. This progression of on-demand viewing and the rise of binge-viewing are the major differences between Netflix and DTV. Nevertheless, while these changes are important, they are not radical; again, as both eras of research show, shifts are gradual.

Interestingly, one of the Sky Digital interviewees envisioned a Netflix-like service back in the early 2000s. When asked to comment on the limitations of Sky Digital and describe his ideal television experience, the discussion progressed as follows:

> V.T.: OK. Do you have a picture of what your ideal TV might be?
>
> RESPONDENT: I think it would be, the way I would like to see it working, right? Sky at the moment got all that programs on the day so you get the broadcast now as it happens. Why can't they record those, put them through the programming and you select what programs you want to watch, the pre-recorded episodes, at the time you want to watch them, when you want to watch them. So why not have a whole week's programming pre-recorded and you can go in and select what you want.
>
> V.T.: Even more options for you.
>
> RESPONDENT: Yeah. That's what I want. Choice effectively. Even with digital TV you are still restricted on choice. They still put a lot of repeats on. I want to watch, for example if I wanted to watch *Die Hard* (1988) that probably wouldn't come on Sky till 9 o'clock. Why can't I watch that at 11 o'clock Saturday morning, if I want to watch *Die Hard* at 11 o'clock on a Saturday morning? Effectively what I have to do is record it myself and then watch it then. Why should I?

Similar points were made by a number of Sky Digital interviewees who, even though they appreciated the channel choice and program variety of the digital service compared to their analog television, still found this supposed emancipation limited in practice. A few dissatisfied users also complained about repeats, signaling that choice after all was controlled and regulated by the broadcaster. Indicatively one male interviewee said: "I think the movie channels and various other channels are very repetitive. Not happy that charges will increase," while another complained, "too many repeats, not enough new innovative comedy, sci-fi, adventure programs." Meanwhile, a 38-year-old subscriber was more direct about his and his family dissatisfaction with the viewing options and cost: "We have had Sky Digital for nearly two years now and we think it is overrated, too expensive, there are far too many repeats. After all we pay more than 100 [pounds] per year to watch ordinary TV and that's all repeats. So why should we pay more each month to watch even more repeats? The only thing not repeated on Sky is the news/weather."

Even with the introduction of features like PPV and more scheduling control, Sky still provided just a partial freedom from normal broadcasting. Users could select from a list of 25 movies offered by the provider with starting times as frequent as every 15 minutes—but for an additional cost. As one male interviewee noted, "Sky Digital is OK but doesn't represent very good value for money as the films are repeated too often and you have to pay extra for special sporting events." Others were even less kind. An unhappy male subscriber commented, "Films, pay as you view, are ridiculously expensive, almost theft. Videos are cheaper. Cannot start to see why they have to charge up to 3 pounds!!! B------s," while a frustrated female customer called Sky Digital "a very good service which is spoiled by very high monthly subscription costs and Pay-Per-View."

Perhaps surprisingly, the bundling of channels on Sky Digital did not meet subscribers' expectations. This system was, in effect, forcing consumers to subscribe to channels they were not interested in just because they were bundled with others that they were—mirroring the cable system in the United States. Of course, this is something that today's Netflix subscribers certainly have no reason to complain about. As one Sky Digital subscriber told me, "The subscriber should be given the chance of what channels he/she wishes to watch and pay Sky accordingly instead of having to purchase packages the majority of which are of no interest." Netflix rectifies this problem, giving users that choice so desperately called for by Sky Digital subscribers. Still, these Sky Digital subscribers were generally infatuated by the prospect of more television content of their choice available, bundled or not. Most DTV subscribers appreciated the enhanced choice offered by Sky Digital compared to "old" television content, but still wanted more—more and better of what they already had, and at a better price. Such change came with Netflix, transferring more power from the producer and content provider to the consumer.

Overall, changes in some media habits and practices co-exist with the endurance and persistence of others—and are usually changes of degree, as the excerpts from respondents imply. The technology surrounding television may be changing quickly, but audience behavior and habits change far slower. Many of today's digital- or web-based media habits spawn from older forms, tastes, preferences, and practices established around the early digital era of television. Meanwhile, though Netflix viewers can now be more focused, selective, and excessive, these habits are adjusted and modified to fit everyday life and domestic practices—just as all viewing practices have been since the introduction of television.

At the time of the DTV research, it was becoming increasingly common for viewers to watch programs alone, signaling a shift in the standard family television viewing experience. Today, the on-demand nature of DVRs and Netflix enables viewers to watch whenever they are in the mood, and across a litany of devices—from screens big and small. The spatial context of viewing has shifted, the traditional schedule is more flexible than ever, and the television set is changing shape, size, and form. Yet, people still watch television programs—whatever the interface. Television has changed, yet programing and content is still what matters most to most viewers.

More than 15 years after the launch of DTV, television is still a significant part of our lives. In the past years, people have equipped their multimedia houses, rooms, and living rooms, built these around DTV—which is connected to the DVR, the Digital Set Top Box/Decoder, the laptop, Netflix and other digital services, etc.—and kept in their pockets or bags portable converged media such as iPhones, iPads, or e-book readers. Digital technology is so embedded in our lives that it will soon no longer make sense to talk about such distinct media technologies or services, without taking into consideration the converged digital environment of which they are all present. However, it is intriguing how audiovisual content still matters, and the viewing of programs remains central to most people around the world today, no matter from which interfaces they access these programs. This resilience of televisual content, its use for entertainment or information purposes, and, of course, its evolution and advancement in production lines with the enhanced techniques and technology used today are worth studying further. So too is the particular impact and attraction of certain successful Netflix's programs on viewers as well as the role of customer analytics and big data in the commissioning, creation, and production of such programs.

NOTES

1. Vivi Theodoropoulou, "The Introduction of Digital Television in the UK: A Study of its Early Audience," dissertation, London School of Economics and Political Science, London, England, 2012, http://etheses.lse.ac.uk/349/.

2. Nathan Rosenberg, *Exploring the Black Box: Technology, Economics, and History* (Cambridge: Cambridge University Press, 1994), 9–10.

3. Ofcom, "Infrastructure Report: Ofcom's Second Full Analysis of the UK's Communications Infrastructure," 2014, accessed January 25, 2015, http://stakeholders.ofcom.org.uk/binaries/research/infrastructure/2014/infrastructure-14.pdf.

4. Everett M. Rogers, *Diffusion of Innovations* (New York: The Free Press, 1962).

5. Elizabeth Evans and Paul McDonald, "Online Distribution of Film and Television in the UK: Behavior, Taste, and Value," in *Connected Viewing: Selling, Streaming and Sharing Media in the Digital Era*, ed. Jennifer Holt and Kevin Sanson (London: Routledge, 2014), 158–180. Having said that and despite the success of online services, it is important to mention that traditional television or "linear broadcasting" as it is now called, remains the most popular way of watching television for most viewers. See Ofcom "Infrastructure Report" and Jennifer Gillan, *Television and New Media: Must-Click TV* (New York: Routledge, 2011).

6. Jennifer Holt and Kevin Sanson, "Mapping Connections," in *Connected Viewing: Selling, Streaming and Sharing Media in the Digital Era*, ed. Jennifer Holt and Kevin Sanson (London: Routledge, 2014), 1–15; Evans and McDonald, "Online Distribution of Film and Television."

7. Raymond Williams, *Television: Technology and Cultural Form* (London: Routledge, 2003).

8. Chuck Tryon, "TV Got Better: Netflix's Original Programming Strategies and Binge Viewing," *Media Industries Journal* 2.2 (2015); Lauren Johnson, "Netflix Is Now a 'Global TV Network' After Launching in 130 New Countries," *Adweek*, January 6, 2016, accessed January 25, 2016, http://www.adweek.com/news/technology/netflix-now-global-tv-network-after-launching-130-new-countries-168862; Jon Lafayette, "Netflix Admits It Faces Challenges," *Broadcasting and Cable*, December 11, 2015, accessed December 14, 2015, http://www.broadcastingcable.com/sites/default/files/public/lead-in.pdf.

9. Elizabeth Evans, *Transmedia Television: Audiences, New Media and Daily Life* (London: Routledge, 2011); Amanda D. Lotz, *The Television Will Be Revolutionized* (New York: New York University Press, 2007); Yu-Kei Tse, "Television's Changing Role in Social Togetherness in the Personalized Online Consumption of Foreign TV," *New Media & Society* 18.8 (2016): 1547–1562; William Uricchio, "TV as Time Machine: Television's Changing Heterochronic Regimes and the Production of History," in *Relocating Television: Television in the Digital Context*, ed. Jostein Gripsrud (London: Routledge, 2010), 27–40; Ofcom, "Infrastructure Report," 2014.

10. Lotz, *The Television Will Be Revolutionized*; William Uricchio, "The Future of a Medium Once Known as Television," in *The YouTube Reader*, ed. Pelle Snickars and Patrick Vonderau (Stockholm: National Library of Sweden, 2009), 24–29; Jenna Bennett, "Introduction: Television as Digital Media," in *Television as Digital Media*, ed. Jenna Bennett and Niki Strange (Durham: Duke University Press, 2011), 1–27.

11. Evans, *Transmedia Television*.

12. As I state elsewhere, "In the late 90s and early years of 2000s, DTV was introduced in the United Kingdom given the planned switch-off of analog television and Europe-wide policies for a total transition to digital broadcasting. It was also launched as an attempt to converge the functions of television with those of the computer, and to potentially bridge digital divides and offer internet-like services and access across the population. In the early days of DTV services, the United Kingdom was considered the most developed market in the world." See Vivi Theodoropoulou, "Convergent Television and 'Audience Participation': The Early Days of Interactive Digital Television in the UK," *VIEW Journal of European Television History and Culture* 3.6 (2014): 69–77.

13. PPV was providing a choice of up to 25 films per night with a frequency/starting time of every 15 minutes.

14. The survey research was conducted using a simple random sample of 1986 Sky digital subscribers. It achieved a response rate of 35.25 percent and a total of 700 responses.

15. Theodoropoulou, "Convergent Television." See also Theodoropoulou, "The Introduction of Digital Television" and Theodoropoulou, "Consumer Convergence: Digital Television and the Early Interactive Audience in the UK," in *Broadcasting and Convergence: New Articulations of the Public Service Remit*, ed. Taisto Hujanen and Gregory F. Lowe (Gothenburg: Nordicom, 2003), 285–297.

16. Theodoropoulou, "Convergent Television."

17. Theodoropoulou, "The Introduction of Digital Television in the UK"; Theodoropoulou, "Consumer Convergence."

18. *Ibid.*

19. By the end of 2001, the United Kingdom saw the fastest DTV penetration in the world, with an overall take-up of 37 percent. See Theodoropoulou "Convergent Television," Theodoropoulou, "The Introduction of Digital Television in the UK," and Theodoropoulou, "Consumer Convergence."

20. Anonymity was ensured for all participants in the research. In the Netflix interview excerpts that follow, the names used are pseudonyms. In the quotes used from the research on DTV, the gender and/or age of the interviewee is provided, in the text, as an identification

mark instead. I would like to thank all the interviewees, participating in both the Netflix and DTV research for their help and valuable insight.

21. Roger Silverstone and Leslie Haddon, "Design and Domestication of Information and Communication Technologies: Technical Change and Everyday Life," in *Communication by Design: The Politics of Information and Communication Technologies*, ed. Robin Mansell and Roger Silverstone (Oxford: Oxford University Press 1996), 44–74.

22. Evans and McDonald, "Online Distribution of Film and Television in the UK."

23. Philip M. Napoli, *Audience Evolution: New Technologies and the Transformation of Media Audiences* (New York: Columbia University Press, 2011), 62.

24. Hye Jin Lee & Mark Andrejevic, "Second-Screen Theory: From the Democratic Surround to the Digital Enclosure," in *Connected Viewing: Selling, Streaming and Sharing Media in the Digital Era*, ed. Jennifer Holt and Kevin Sanson (London: Routledge, 2014), 40–61.

25. Roger Silverstone, Eric Hirsch, and David Morley, "Information and Communication Technologies and the Moral Economy of the Household," in *Consuming Technologies: Media and Information in Domestic Spaces*, ed. Roger Silverstone and Eric Hirsch (London: Routledge, 1992), 15–31.

26. Uricchio, "TV as Time Machine," 35.

27. Tse, "Television's Changing Role in Social Togetherness," 1550.

28. But, apart from the audience's or consumption side, this goes, more perhaps, for the development and design of new media technologies and offerings as well. Meikle and Young note how new convergent media are based on longer historical trajectories. See Graham Meikle and Sherman Young, *Media Convergence: Networked Digital Media in Everyday Life* (New York: Palgrave Macmillan, 2012).

29. Wendy Van den Broeck, Jo Pierson, and Bram Lievens, "Video-on-Demand: Towards New Viewing Practices," *Observatorio Journal* 3 (2007): 39.

30. Lisa Gitelman, *Always Already New: Media, History, and the Data of Culture* (Cambridge: MIT Press, 2006), 6.

31. Philip Palmgreen and Jay D. Rayburn, "An Expectancy-Value Approach to Media Gratification," in *Media Gratifications Research: Current Perspectives*, ed. Karl E. Rosengren, Lawrence A. Wenner, and Philip Palmgreen (Beverly Hills: Sage, 1985), 61–72; Terje Rasmussen, "New Media Change: Sociological Approaches to the Study of New Media," in *Interactive Television: TV of the Future or the Future of TV?* ed. Jens F. Jensen and Cathy Toscan (Aalborg: Aalborg University Press, 1999), 161.

32. Tryon, "TV Got Better"; Evans and McDonald, "Online Distribution of Film and Television in the UK"; Holt and Sanson, "Mapping Connections"; Napoli, *Audience Evolution*; and Debra Ramsay, "Confessions of a Binge Watcher," *Critical Studies on Television*, October 4, 2013, accessed July 23, 2015, http://cstonline.tv/confessions-of-a-binge-watcher; William Proctor, "It's Not TV, It's Netflix," *Critical Studies on Television*, November 29, 2013, accessed July 23, 2015, http://cstonline.tv/netflix.

33. Ramsay, "Confessions of a Binge Watcher."

34. Tryon, "TV Got Better," 106.

35. Ramsay, "Confessions of a Binge Watcher."

36. Tryon, "TV Got Better," 112.

37. Theodoropoulou, "Consumer Convergence," 295.

38. Theodoropoulou, "Convergent Television"; Phillip Swann, *TV Dot Com: The Future of Interactive Television* (New York: TV Books, 2000); Michael A. Noll, "TV Over the Internet: Technological Challenges," in *Internet Television*, ed. Jo Groebel and Darcy Gerbarg (Mahwah, NJ: Lawrence Erlbaum, 2004), 19–29.

39. Theodoropoulou, "Convergent Television," 74.

40. Lotz, *The Television Will Be Revolutionized*.

Digital Delivery in Mexico
A Global Newcomer Stirs the Local Giants

Elia Margarita Cornelio-Marí

When Netflix launched in Mexico in September 2011, it seemed like a mere curiosity—attractive to a minority of premium users who had access to broadband Internet at home, and those few that were willing to pay for media content instead of downloading illegally. At the outset, the library offerings were so limited (e.g., old Hollywood movies, even older second-rate local films, some Latin American telenovelas) that many early adopters chose not to continue with the service after the one-month free trial. Eventually, the availability on diverse video game and mobile platforms, new offers in recent content as well as the integration with Facebook, contributed to Netflix's growing popularity in Mexico. During its first four years of expansion, Netflix has become a popular option for films and television series catered to the tastes of the local audiences that have access to broadband, to the point that in June 2015 it holds 55.7 percent (out of four million users) of the over-the-top (OTT) market.[1] Nevertheless, the company's market share has been rapidly slipping—down from 70 percent in 2014—because local media companies have been observing closely, copying its business model, and employing their strongest resources in order to compete with the streaming video giant.[2]

Mexico is home of a powerful media and telecommunications industry, which historically has extended its business to other Latin American countries and to the Hispanic population of the United States.[3] The country is the domestic market of Grupo Televisa, the largest media company in Latin America as well as of the telecommunications behemoth América Móvil (number 155 in the Fortune 500 in 2015).[4] The entry into and success of Netflix in Mexico has established a new market that now seems very attractive to many significant local firms. The availability of titles offered by Netflix has

201

already affected access cycles to many American television series and movies, threatening cable and satellite television providers. More recently, with the launch of exclusive series (*Camelia La Texana* [2014]), exclusive local films (*La Dictadura Perfecta* [2014]), and its original production for Mexico and Latin America (*Club de Cuervos* [2015], *Narcos* [2015–], *3%* [2016–]) Netflix is directly reshaping the hegemony of the established local producers. These producers, in turn, are deploying all of their respective structural advantages to gain a foothold into the video-on-demand sector, not only in the country but also in the Latin American region at large.

The arrival and evolution of Netflix in Mexico offers a privileged case to analyze a set of interconnected issues related more generally to the growing spread of technologies for digital distribution of film and television around the world, namely:

- The technological, regulatory, and logistical challenges that the developing local markets impose to global players;
- How local culture acts as a restraining factor that requires global players to adapt in order to thrive in the new market; and
- How the entry of a global player might change the local media industry, including the reaction of local incumbents to the arrival of a new distribution model.

In *On-Demand Culture*, Chuck Tryon states clearly, "Given the global scope of the entertainment industry, it is crucial to recognize the intersections of the local and the global when it comes to digital delivery. This analysis requires careful attention to a variety of political, logistical, and cultural factors that influence how audiences access movies [and television] across a range of countries."[5] This essay answers this call while also acknowledging that these analyses must place themselves within an industrial framework that inevitably has to take into account cultural aspects in order to offer a complete vision of the impact of the new model of distribution that Netflix has heralded in Mexico.

Methodologically, I rely on the coverage of Netflix's activity by the local and international press, reports to investors, and statistics by media consulting firms, as well as on direct analysis of the platforms and advertising strategies of the local competitors. On the theoretical level, I make use of recent literature on digital distribution and studies related to the international spread of television, which help explaining the dynamics—cultural and otherwise—that emerge when global distributors reach local markets and get in contact with their incumbents and audiences.

The Latin American Opportunity

Repeatedly, during the writing of this analysis, it will prove essential to look at the larger framework of Latin America in order to understand Netflix's role in Mexico. The reasons behind this are the evident structural similarities in terms of the Internet access and economic development that Mexico shares with all the countries in the region, the status of the Spanish-speaking Latin America as a geolinguistic region for the consumption of media, and Mexico's leading role as a media producer, adaptor, and distributor for this territory.

After entering into Canada in 2010, Netflix decided to expand to Latin America and the Caribbean during 2011, rolling into 43 countries in the region.[6] Surprisingly, the company chose a territory comprised of developing countries as the second move in its plans for international expansion. According to Netflix leadership, the reasoning behind this choice was the great opportunity that the region represented in long-term growth. "[G]iven that Latin America has about 4x more broadband households than Canada, there is lots of room for growth," stated the letter to the shareholders from the third quarter of 2011.[7] For the launch, *The Wall Street Journal* consulted Credit Suisse Senior Analyst John Blackledge on the matter. He pointed out that "the new region has a sizable chuck of potential customers. Brazil, Mexico, and Argentina alone have nearly 35 million broadband subscribers ... compared with 10 million in Canada, where Netflix began its international expansion."[8] Unlike Apple, whose foray into Latin America had been quite slow relative to its expansion in other global regions—the iTunes store did not open in Mexico and Brazil until 2009 and 2011 respectively—Netflix clearly wanted to score the first-mover advantage and establish itself in a market where legal video streaming services were uncommon, or even nonexistent.

Of course, Netflix's decision to enter into the Latin American market was aided by the sizeable economic prospects in the region. However, while the region appeared like virgin territory, it was not an Eden; there were unusual challenges that the company faced very quickly. An article in *The Huffington Post* noted:

> The move brings challenges not seen in Netflix's core markets, the U.S. and Canada. Broadband Internet reaches a far smaller percentage of homes in Latin America than in the United States, and speeds are slower. Piracy of movies is among the most widespread on the planet, meaning many consumers can pick up a DVD or CD of the latest films for less than a dollar. Also, Netflix has little brand recognition in the region, and in the case of Brazil it already faces a homegrown competitor [i.e., Net Movies].[9]

The Mexican market also presented most of these discouraging traits, but on the bright side the country offered a enormous growth potential in Internet use, an audience hungry for Hollywood film and television, and an untapped non-regulated video streaming sector with practically no local competition. All of these factors created a challenging, yet very attractive new territory for a global media player.

Factors Surrounding the Entry and Evolution of Netflix in Mexico

Netflix launched its streaming service in Mexico in September 2011. The event was closely followed by the main national newspapers. *El Universal* explained the service in painstaking detail; *El Economista* called it a "digital revolution in content" and published an info-graphic illustrating the service's functions; and *Reforma* decided to focus on Netflix's CEO Reed Hastings statement that the service would not compete with television.[10] Netflix was presented as a curiosity that could attract the attention mostly of tech enthusiasts. Netflix was not the absolute first video-on-demand product in Mexico—the iTunes store added video rentals soon after its Mexican launch in 2009—but it was the first platform to offer unlimited streaming access for a single monthly fee.

The following pages present a detailed analysis of the multiple factors— both facilitating and deterring—that have surrounded the entry and evolution of Netflix in Mexico, including the major obstacles the service has encountered and a brief account on how it has faced them so far. For clarity's sake, I have decided to organize these factors in three broad areas, each of them linked to one of the issues that this essay addresses: (1) technological, regulatory, and logistical challenges; (2) local culture as a restraining factor; and (3) reaction of the local incumbents. It should be noted, though, that this is an artificial separation, for all of these factors are, in reality, interconnected and affect each other deeply. Drawing from press sources, state statistics and regulations, and research on the role of local culture in international distribution, I examine Netflix's entry into Mexico.

Technological, Regulatory and Logistical Challenges

This first section details the features of the Mexican market at the time of Netflix's arrival, including the technological infrastructure and legal framework that would eventually regulate the company's practices. It also addresses billing and pricing, key considerations Netflix had to adjust in order to remain viable in Mexico.

Technology

The first of the challenges that Netflix faced in Mexico was the low penetration of broadband Internet, which is directly correlated to the market's lack of competition in telecommunications. Indeed, Telmex (a subsidiary of América Móvil) holds a de facto monopoly in many regions of the country on landline telephone and broadband Internet. Telmex's market dominance and its reluctance to provide higher speeds at affordable prices prevented expansive Internet penetration in Mexico—to such an extent that it placed near the bottom in broadband use among the 35 member countries of the Organization of Economic Co-operation and Development.[11] Statistics from ITU and the World Bank claim that only 12 percent of the total population has access to fixed broadband Internet. Meanwhile, mobile broadband penetration had only reached 37.5 percent of the population by 2015.[12] Basic access to the Internet is a bit more encouraging, as 44.4 percent of Mexicans were regular users in 2014.[13] Although there is still little broadband penetration, there is notable growth year to year. For instance, the World Bank reports that in 2013 broadband access was 10.93 percent, while a year later it was 11.56 percent; similarly, the number of Internet users grew slightly from 43.5 percent in 2013 to 44.4 percent in 2014. This is a clear expansion in both cases. In summary, the country is still far behind in comparison to developed nations, but growing at a steady rate, a factor that surely encouraged Netflix's decision to pursue this new market.

Legal Framework

In Mexico, the Congress of the Union regulates the action of the telecommunications and broadcasting companies. In 2011, there were separate rules for each sector in place, as they were at the time considered as two different industries. While Televisa dominated broadcasting, América Móvil was the leader in telecommunications. With such regulation in place, América Móvil was prohibited to offer pay television and Televisa could not offer telephone or Internet service. Of course, new developments in digital technology inspired the creation of a new regulatory environment. After long debate, the new Ley Federal de Telecomunicaciones y Radiodifusión was passed in July 2014, with the primary goal of enhancing competition in what is to be a new, growing sector. As a result of these changes, more steady growth in Internet penetration has already begun throughout the country, which in turn is likely to create an OTT-friendly environment that could benefit Netflix—but which also brings with it a pair of formidable local competitors.

Nevertheless, Netflix benefitted significantly from an absence in the preexisting legislation when it launched in Mexico. Though the country places limitations on foreign investment in broadcasting, Netflix was exempt from these limitations as a video-on-demand/Internet company. As a result, despite its status as a foreign firm, Netflix faces no taxes and few legal limits in its new territory.[14] These circumstances have certainly allowed the company to act more freely in a moment of increasing deregulation in Mexico.

Billing

Originally, Netflix's billing was dependent on the use of credit cards, which are not as widespread in Mexico as they are in the United States and Canada.[15] Debit cards are more popular, but the local banks do not authorize use over the Internet in all cases due to lingering distrust of online transactions and a perceived lack of punishment for cybercrime. This became a real problem for Netflix in Latin America, as demonstrated in investor letters from 2011 and 2012. For instance, the report for the fourth quarter of 2011, soon after Netflix's entry to the region, mentioned the issue of billing quite superficially:

> As expected upon launch, we've found that processing ecommerce consumer payments is quite challenging as compared with North America and Europe. To overcome this challenge, we are working with our local payment partners to optimize our systems, exploring adding new payment methods and testing various trial campaigns to improve conversion.[16]

Comparatively, the report from the third quarter of 2012 was much more explicit on the issue:

> The biggest issue holding back much stronger growth is payments. For a variety of reasons, many Latin American broadband households are leery of, or unable to, provide debit/credit cards that can be accepted over the Internet. For those who do provide us debit/credit cards, we see higher rates of payment declines than in our other markets.[17]

Netflix first addressed the billing problem in Mexico by promoting debit cards as an accepted form of payment. The letter to shareholders from the fourth quarter of 2012, explains, "In Mexico, we've made progress in enabling debit cards for Netflix which has to be done in some cases bank-by-bank."[18] Secondly, Netflix followed the strategy of many other companies like Apple, Microsoft, and Nintendo in launching prepaid cards to access online services. On November 2014, Netflix made available gift cards for 99 and 299 pesos, which could be acquired in supermarkets and department and convenience stores.[19] For Netflix, this was an answer to the specific conditions of the market. Mexican consumers are used to prepaid cards as a way to control their expenses and protect their credit/debit cards from online fraud.

Pricing

Netflix thrives on the idea that users want to pay an affordable monthly fee to get a high volume of valuable content. In Mexico, the service decided to enter with a fixed price of 99 pesos a month, which would seem decidedly cheap for the middle class users, who were paying already up to three times that amount for expensive broadband access and up to five times that for premium direct-to-home television subscriptions.[20] In October 2014, Netflix introduced a three-tiered pricing scheme in Mexico: $89 for one streaming device with standard definition, $109 for the basic service of two streaming devices on high definition, and $149 for up to four streaming devices in HD. In a gracious move, Netflix decided to keep the basic service at $99 for the existing clients until October 2016 when the prices would be set at $99 for one screen, $129 for two screens, and $159 for four screens.

Price is closely linked to perceived value. In Mexico, Netflix offered value to consumers based on perceived flexibility and improved content. From the initial point of entry until a price adjustment in 2014, there was no limit on the number of devices that could be used at the same time on the same account. This flexibility fostered sharing between extended families and friends, in turn providing Netflix with huge amounts of data for tracking viewing habits. Furthermore, this practice created a positive image of the company, ensuring loyalty and the perception of value sorely needed to gain a foothold against piracy. Likewise, as in the United States, Netflix dangled a one-month free trial to users. This was, unsurprisingly, a very popular method of luring in new users, especially as the company added more and more original content to lure users into paying clients.

Local Culture as a Restraining Factor

Experienced media companies know that local culture is a force to reckon with. Despite the claims of global homogenization, local audiences still prefer watching content that feels closer to their own ways of life. Netflix faced this reality when it entered the Latin American region. This second area of analysis comprises three linked topics: (1) the localization of content; (2) characteristics of the audiences for Netflix in Mexico; and (3) the strategies the service has put in place to raise its brand awareness.

Localization of Content

Content has become a crucial factor to explain both the early difficulties and the recent successes of Netflix in Mexico. Being clearly connected to culture, content could work both as a deterrent to entry for global firms and as a powerful enabler, once the question of cultural tastes has been adequately addressed. During the first wave of diffusion of satellite television networks in Europe, Jean Chalaby proposed a model to explain the strategies that established global media companies use to succeed in markets characterized by strong cultural identity.[21] The model identified four steps that the Pan-European channels were putting in practice. These steps move from lower to higher levels of "localization": (1) introduction of local advertising, (2) translation of content, (3) local programming, and (4) local opt-out. The first step cannot be applied to Netflix because it works on a subscription basis, but the other three steps accurately describe how the company operates in Latin America, albeit in different order: (1) translation, (2) local opt-out (separate feed for each country, easily doable in the case of streaming technology), and (3) local programming. In this section, I reflect on the form that each of these steps is taking within Mexico.

TRANSLATION

The first step of Netflix's adaptation process in Mexico was obviously translation: subtitling and dubbing. In the country, the preference for one method of language transfer over another depends on age (kids programming must be dubbed) and genre (animation and documentaries are preferably dubbed; drama could be either dubbed or subtitled). In addition, it depends on the characteristics of the perceived audience, as viewers that are purportedly more sophisticated prefer subtitling or to even watch programming in the original language. This preference for subtitles, which is clear in the middle and higher classes that constitute Netflix's target audience in Mexico, apparently has come as a surprise for the company. As a result, it had to rush to offer both methods of language transfer. In the first quarter of 2012, the company reported to have "increased … subtitle coverage [in Latin America] to nearly 100 percent of non-kids English language content (in addition to previously available dubbing) to accommodate varying viewing preferences."[22]

Regarding dubbing, Latin America has historically developed a unique linguistic dialect called Español Latino that has become widely present throughout the region. This variety of the language, which differs substantially in vocabulary and pronunciation from Castilian Spanish, is built upon the idea of a "neutral" pronunciation and vocabulary understandable in the Americas.[23] It is not actually spoken anywhere, but people accept it as a normative language for watching films and television. It has developed slowly through time, to the point that it has become a convention used in the major dubbing studios in Mexico, Argentina, Chile, Colombia, and Venezuela. Creating a dubbed feed that works for several countries surely helped limit Netflix's expansion costs. For its original English productions, Netflix became a commissioner to dubbing studios in the region—*House of Cards* (2013–) is dubbed in Mexico, *Orange Is the New Black* (2013–) in Chile, and *Bloodline* (2015–2017) in Argentina.[24] For the rest of the American content, the dubbing and subtitling is unchanged from the initial distribution window (be it a theatrically released film or pay cable television series). Despite early struggles, the establishment of these two language options heralded few complaints from the audience (with the exception of some anime fans demanding that the default track be Japanese, not an English dubbing.)[25]

LOCAL OUTPUT

Netflix was aware that in order to gain acceptance in Mexico, it had to do more than translate its interface to Spanish language—it had to become more local in terms of content. Internet protocols are geographically sensitive. This enables Netflix to divide its feed and tailor the content for each country. Indeed, Mexico has its own Netflix feed, which features content that is not available in other countries of the region, and vice versa. This differentiation responds both to diverse tastes and to the "need to buy" rights for each country. Moreover, it is generally accepted that Mexico's geographic proximity to the United States means that Mexican audiences have a taste for American film and television. More than 80 percent of movie theater screens present imported fare, most of it coming from the United States. American series (both comedy and drama) constitute a substantial part of the schedule in television channels devoted to entertainment as well as on cable and satellite systems.[26] In spite of this visible preference for Hollywood productions, most Mexicans would prefer to watch national newscasts, sports, reality shows, comedy series, and telenovelas on the leading broadcast networks. Telenovelas are a very well established genre that garners large followings and the highest shares of overall television ratings.[27]

The preference for local content over foreign content of the same type and quality has been noted by media scholars, and theorized as "cultural proximity."[28] Generally, local audiences choose films and television programs that spotlight situations and characters closer to what they experience in their daily life, and not only that—they also want them to be told in their traditional, familiar narrative forms. Audiences apply a "cultural discount," implying that they reject audiovisual products that are too distant to their own way of life.[29]

In terms of taste, scholars have considered Spanish-speaking Latin America as a "geolinguistic region" with certain shared characteristics such as language, ethnicity, history, and ways of life.[30] As such, Netflix was forced to include some local, or at least Latin American, film and television beginning with its launch in Mexico. At the regional level, Netflix announced a deal with the NBC Universal's Spanish-language broadcaster Telemundo in

July 2011, which would provide 1,200 hours of programming per year, including hit series like *La Reina del Sur* (2011).[31] In August 2011, Netflix made similar agreements with the Argentinean Telefé and the Colombian RCN, producer of the international success *Yo soy Betty la Fea* (1999–2001).[32]

In Mexico, the acquisition of local catalog content was achieved very early, by signing deals with such producers as Televisa and TV Azteca. These agreements were made public in May 2011, in preparation to the launch of the service in the country. Televisa would give Netflix non-exclusive access to 3,000 hours of catalog telenovelas and series a year after they had been broadcasted, while TV Azteca would provide another 1,000 to 1,500 hours of programming.[33] Additionally, Netflix signed a contract with the Mexican indie studio Canana, producer of the successful series *Yo soy tu fan* (2010–2012), which previously aired on Mexican public television.[34] With these deals in place, Netflix obtained access to a vast catalog of the most successful Spanish language telenovelas earning an excellent position to cater to the tastes of its new audiences.

Another step of Netflix's adaptation to the Mexican market was executed in August 2014, when Netflix started offering an exclusive telenovela, *Camelia La Texana*, a narco story produced by the independent studio Argos.[35] In the United States, this telenovela aired on Telemundo, but it failed to reach Mexican screens either on pay or on free-to-air television. For Netflix, it constituted a first attempt at product differentiation for the region, in its ongoing effort to lose the label of a mere aggregator of old titles. Meanwhile in 2014, Netflix acquired a slate of high-profile Mexican films like *Cásese quien pueda* (2014), *Cansada de besar sapos* (2006), and *No se aceptan devoluciones* (2013). Likewise, in March 2015 Netflix began streaming the political satire *La Dictadura Perfecta*, the highest grossing Mexican film of 2014, which, incidentally, parodies the relationship between the federal government and Televisa.

Offering exclusive local content was a logical move for Netflix, but in doing so, the company announced itself as a direct competitor to the established telenovela outlets, particularly Televisa and TV Azteca—both of which provided significant amounts of content to Netflix in Latin America. The breakup with Televisa would come after four years of collaboration. Following the launch of its second OTT service (Blim), Televisa announced that it would withdraw all of its content from Netflix in the last quarter of 2016, in order to strengthen its own strategy of cultural proximity.[36]

PRODUCTION OF LOCAL PROGRAMMING

The final step in the model of localization adapted from Chalaby is the production of original local content. Netflix positioned itself as a new distribution channel for local producers, but later it became a direct competitor for viewer attention, and eventually grew into a desired outlet for producers hoping to reach those viewers. On April 2014, Netflix announced the production of comedy series *Club de Cuervos*, created by Gaz Alazraki, the director behind the most successful Mexican film of recent years: *Nosotros los Nobles* (2013). The 13-episode Spanish language series was shot entirely in Mexico with an international cast from Latin America and Spain. It launched in August 2015, and renewed for a second season in October of the same year. The story revolves around a family that owns a soccer team, and while a dramedy, it offers clear melodramatic characteristics and even includes some classic telenovela subplots. Much of the humor stems from the sardonic depiction of the world of soccer, thus profiting from the high interest of Latin Americans for everything that has to do with the sport.[37] Almost simultaneously in August 2015, Netflix launched *Narcos*, a series based on the life of the notorious Colombian drug dealer Pablo Escobar, starring Wagner Moura and directed by Brazilian José Padilha.[38] This production is partly spoken in Spanish and includes an international cast of actors from Chile, Colombia, Mexico, and the United States. Finally, for 2016, Netflix announced new original productions based in Latin America: the Mexican political drama *Ingobernable* (2016–), starring Kate del Castillo, and the Brazilian dystopian thriller *3%*, to be spoken in Portuguese, which deals with the issue of social inequality.[39]

The significance of Netflix's original production in Latin American lies in its regional scope. It builds on the existence of a geolinguistic region that covers also the Hispanic population of the United States, thus ensuring large economies of scale for the company. Just as the Latin American media companies have been doing for decades, Netflix is producing *Club de Cuervos* and *Ingobernable* in Mexico, *Narcos* in Colombia, and *3%* in Brazil knowing that these programs will be watched throughout the continent and, with some luck, around the world. In this sense, Netflix is very wisely creating series based on topics—crime, family, social inequality, and soccer—preferred by audiences of the region but which also have universal appeal.

The Importance of Localization

Despite its age, Chalaby's model can be applied with some adaptations to study the entry of Netflix in Latin America; the case is a fundamental example of a global firm adapting to a regional marketplace. The general strategy for Netflix in the country and the region seems to be focused on increasing the cultural proximity of its offerings; at the same time, it continues to provide international and Hollywood content. This mix of local and global content is a general practice of Netflix in international markets. Ted Sarandos, Netflix's Chief Content Officer, explained that programming is "around 15 percent to 20 percent local ... with the 80 percent, 85 percent being either Hollywood or other international content. One of our first indicators that we are getting the mix right is how many hours of viewing people are participating in."[40] For Mexico, the mix seems to have been compelling enough, since in 2012 Mexicans were reported to be "watching more Netflix than their counterparts in the US, Canada, or the UK."[41]

From this data comes a picture in which content has proven crucial for a successful regional expansion. Be it in the form of careful selection of catalog content or the production of new original series, local/regional content is a key competitive advantage for Netflix in Latin America, just as the original series are for its domestic market. As I stated at the beginning of the section, culture could act both as a deterrent and as an enabler for the entry and evolution of a global player such as Netflix. Local culture is an initial deterrent because it demands adaptation, high initial investment in language translation (subtitling and dubbing), and the acquisition of catalog content that should follow local preferences in genre (e.g., telenovelas) and theme (e.g., soccer, narco stories). However, for a company like Netflix that employs sophisticated tracking tools used to identify the tastes of the viewers, culture can rapidly become a catalyst for further growth.

First, the existence of a unique language and shared cultural traits opens an opportunity to create content that could resonate in the whole region. In this sense, the decision to produce *Club de Cuervos* in Mexico comes from the awareness of the country's standing as a traditional producer of endearing comedy series (e.g., *El Chavo del Ocho* [1972], *El Chapulín Colorado* [1972]). *Narcos*, on the other hand, builds on the recent tradition of hugely successful crime telenovelas from South America (e.g., Colombia's *El Cartel de los Sapos* [2008]). After the initial learning curve, Netflix seems to have reached a level of knowledge about Latin American audiences that took decades for the regional leaders to build. The success of its Spanish-language programs, which were quickly renewed for second seasons, seems to prove this point.

Secondly, shared language and culture bring opportunities for distribution of series that are absent from the traditional television outlets either because they are considered too foreign (e.g., Spain's *El Tiempo entre Costuras* [2013]) or too old (e.g., Colombia's hit telenovelas *Café con Aroma de Mujer* [1993] and *Yo soy Betty la fea*). There are cultural implications stemming from the availability of this content simultaneously for the entirety of the Latin American region. It demonstrates the diminishing gatekeeping abilities of the local media incumbents, while on the other hand it raises questions about the changes in preferences and practices of the audiences.

Audiences

It should be noted that Mexico is a country with profound inequalities, so that Netflix is a service aimed at local elites: people from middle and upper classes who can pay for broadband Internet at home and for enabled devices (personal computers, smart television sets, video game consoles, tablets, and smartphones). Every piece of analysis regarding the changing habits brought about by systems of digital distribution in Mexico has to consider these social structures. In their recent work on the responses of traditional telecom operators to the entry of OTTs in Latin America, Juan José Ganuza and María Viecens assert, "In the short to medium term, it is expected that in LATAM, the OTT content that requires the high speed access networks will be limited only to reduced areas with consumer segments with high purchasing power."[42] Sinclair makes a similar point when discussing the two largest Latin American media companies. He writes, "The base of both Televisa and Globo remains their pre-eminence in free-to-air domestic markets in countries marked by severe income inequalities between the globalised elites and the common people, la *gente corriente*, who just love their telenovelas."[43] Hence, Netflix's audiences have less to do with the massive audiences of the free-to-air national networks, and more to do with the clients of satellite television and cable systems as well as with the consumers of online piracy.

The next venue of exploration for Netflix in Mexico should be related to reception: who uses it, and for what? What are the new viewing practices? How has it altered the audience's experiences of seriality in television (especially where melodramas are concerned)? However, at of this moment there is a lack of official sources on the matter since the company keeps all of its data confidential. Likewise, scholarship is equally undeveloped on this topic. Instead, I offer some personal observations that might point to general trends.

I was an early adopter myself, opening an account on launch day. I recommended it then to colleagues, who tried it but became disillusioned before the end of the one-month trial because they thought that the offerings were both scarce and aged (they were otherwise well served by pirate streaming and illegal downloading). In the four years since, I have received scattered accounts of Netflix use during informal conversation. I have heard of kids' fascination with the huge availability of cartoons and Disney series, of housewives binge-watching old classic Mexican telenovelas and the Spanish *El Tiempo entre Costuras* (2013–2014), and of a couple (both older than 70) staying awake well into the night watching movies and series. I have seen my own parents enthusiastically learn how to use the interfaces of their Nintendo Wii and Apple TV to watch *The Nanny* (1993–1999), *Café con Aroma de Mujer*, and Cantinflas's films from the 1960s. Nevertheless, I also know of a friend who had both Apple TV and Netflix but had to cancel the latter because he moved to an area not yet served by an Internet or cell phone signal. Finally, I have heard of high-income acquaintances that do not even know what Netflix is, because they are subscribers of premium satellite television (Sky). These disparate stories nonetheless suggest wider changes in the ways that Mexicans now conceive of and partake in access to television and film.

Brand Awareness

In the United States, Netflix has established its brand over an extended period, with DVD rentals and later as a distributor of streaming video content. However, in Mexico, Netflix was a completely new brand, not connected with the local entertainment imaginary. Given that only a minority of Mexicans has access to broadband, services such as Netflix still seem like a novelty to the general population. The company identified this reality very early. The letter to shareholders from the first quarter of 2012 explained that "while there is limited current OTT streaming competition in the region … this lack of OTT competition means that the concept of on-demand streaming video (outside of piracy and YouTube) is nascent, requiring us to do more work in driving consumer understanding and acceptance of our streaming service."[44] In order to increase brand awareness in Mexico, Netflix launched in 2014 the advertising campaign "Vive Netflix" (live Netflix), evoking the nationalist phrase "¡Viva México!" The campaign consisted of three spots: *Oribe*, *Abuelita*, and *Autobús*. This campaign provides a glimpse into how Netflix presents itself to viewers outside of the domestic market and reveals which cultural traits it activates in order to become accepted in the new environment.

1. *Oribe*: Launched just after the end of the 2014 World Cup in Brazil, this spot features Oribe Peralta, a star of Mexico's National Soccer Team. The spot plays with the conventions of thrillers. The popular player gives his wife a scare because he arrives home "too early." When the wife notices his low spirits she cheers him up with the news that the second season of *Orange Is the New Black* is available on Netflix. This advertisement empathizes with the nation's sadness after being disqualified from the World Cup in the heartbreaking match against the Netherlands; it also portrays a beloved soccer figure and at the same time points to both the original content and accessibility of the platform.

2. *Abuelita* One and Two: The first version of this spot is a wink to film fans. The camera follows a little boy on a tricycle that approaches the figure of an old woman sitting on an armchair in a clever spoof of both *The Shining* (1980) and *Psycho* (1960). When the point of view changes, the figure is revealed to be a sweet old woman, wearing high-end headphones connected to a tablet. She asks the boy if he wants to see cartoons

and sits him on her lap so they can watch together. The protagonist of the spot contradicts the stereotyped image of Mexican grandmothers watching telenovelas on free-to-air television, while it simultaneously shows the ease of use of the Netflix platform for elderly people and children alike. The second version of the spot repeats the introductory part with the child approaching the rocking chair from behind; however, this time, the grandmother appears to be sleeping (with earbuds on) and does not react when the boy tries to wake her up. The kid runs away, thinking that she is dead—and only then does the grandmother wakes up smiling to continue watching *House of Cards*. Again, this second version humorously promotes Netflix's original content and the use of the platform by the elderly.

3. *Autobus*: This spot focuses on mobility as a characteristic of digital delivery. The protagonist is a boy riding an urban bus and is exchanging glances and smiles with a beautiful girl. The boy gets off the bus but turns around suddenly, screaming: "Wait!" Then he starts running after the bus. The girl thinks that he wants to talk to her, but soon realizes that he has forgotten his cell phone, which is streaming a series. When the boy finally gets on the bus again, the girl is holding the phone in her hand. So, they both start watching Netflix together. The main idea behind this specific ad is affordability, emphasizing that young people that take the bus can actually pay for a monthly subscription.

The advertising campaign "Vive Netflix" retains the clever references to conventions of Hollywood genres (romance, thriller, and horror) that the brand applied to its Thanksgiving 2014 American ad (*Airport*), and on its Canadian campaign "You gotta get it, to get it" (with ads like *The Proposal, Test Results*). Yet, it also shows local faces in Mexican environments, introducing to the Mexican public the benefits of the system and highlighting key experiential attributes: instant access, ease of use, and affordability.

Netflix implemented another strategy to increase awareness in early 2015. Following the model of a previous contest run in 2014 in Brazil, the firm organized the Netflix Prize, which invited viewers to vote on social network sites for an independent Mexican film that would deserve to be part of its global catalog. The initiative drew the participation of 9,454 Twitter users and 21,431 Facebook users, who decided to award the 2013 film *12 Segundos*. Through this process, Netflix promotes local talent while simultaneously obtains essential data that offers insight about the preferences of Mexican audiences.

Reaction of the Local Incumbents

This last area of analysis focuses on how the local companies have reacted to the entry of Netflix in Mexico. It takes into account both the illegal and the legal competition. Therefore, I start by exploring piracy, and conclude with an overview of the local companies that have decided to enter so far into the subscription video-on-demand (SVOD) market.

Piracy as Illegal Competition

In a November 2014 visit to Mexico, Hastings stated that piracy remains a key competitor for Netflix.[45] This remains true for a variety of reasons. First, at the most basic level, piracy keeps users away from the service altogether. The pervasiveness of piracy has made Mexican people accustomed to obtaining easy access to new releases. Most films are available very cheaply on DVDs sold on the streets, and many Mexicans find it easy to watch content directly on illegal streaming sites like PelículasID and Cuevana2—where they can find recently released films as well as complete seasons of the most successful television series. In spite of the legal provisions against piracy, there is rarely any actual punishment or prosecution for individual consumers. Consequently, piracy is widespread and considered normal, even for members of middle and higher social classes. This attitude reflects widespread conceptions about file sharing at the international level identified by Mahalia Jackman and Troy Lorde in their article about the psychological, social, and economic factors influencing digital piracy. They explain that piracy is not considered a criminal act, but a victimless petty offense, in part because of the detachment that the Internet grants to the lawbreakers.[46] The second reason that piracy remains a threat to Netflix is because it forces the company to include more recent—and thus more expensive—fare. In fact, one of the first objections against the service was "why pay for old movies when you can find new ones on the Internet for free?"[47] As a result, Netflix has strived to keep its library updated, more localized, and more culturally relevant.

It seems that Netflix realized this issue early on because alongside the launch of local content, it started offering some new releases. Particularly salient was the advertising campaign for the launch of *The Hunger Games: Catching Fire* (2013) in Mexico, where it was available for streaming on the same day of the DVD launch, and five months earlier than it was available in the United States.[48] Later, Netflix made the film *The Butler* (2013) available while it was still playing in movie theaters. With these kinds of moves, devised specifically to appease the impatient local viewers, the company worked to change the idea that it was only a place to watch old movies and television. In truth, Netflix just adhered to the same strategy of early premiere that Hollywood studios have long been applying in Mexico for highly anticipated films: it made them available earlier that in the United States in order to beat the pirates. These moves offer clear evidence that Netflix operates differently in different spaces; in this case, adjusting to a market where copyright is not strictly enforced and where people were already accustomed to having immediate access to the newest content.

Finally, piracy has directly victimized Netflix by spreading its original productions illegally. While Netflix original series *House of Cards, Orange Is the New Black*, and *Marco Polo* (2014–2016) cannot be accessed through the local competitors, they are accessible on a variety of the region's most popular pirate streaming websites, such as cuevana2.tv, miratuserie.tv, seriales.us, and seriespepito.com. Sadly, piracy is an issue already very much a part of the Mexican viewing culture. Netflix has little ability to combat this cultural habit beyond promoting its brand and original series, and keeping its prices affordable.

Mexican Competitors for Subscription Video On-Demand

When Netflix arrived in Mexico, it was the first streaming service with a monthly fee, and it opened the market for that kind of service. According to press reports, in the months following Netflix's entry, several similar services sprouted. Currently, the legal competition for Netflix in Mexico is growing quickly, and includes both local and global companies.[49] A complete list of firms connected to digital delivery of television and film in Mexican territory should include providers of pay-per-view over the Internet (e.g., iTunes, PlayStation Video), TV Everywhere systems (e.g., Dish Móvil, Fox Play, HBO Go, Max Go, Blue to Go, Video Everywhere) and advertising-based systems like Crackle. However, for the purposes of this essay, I limited my analysis to Mexican companies that offer subscription plans—either as standalone services or mixed with another model such as PPV or TV Everywhere. This reduces the list to the six major competitors, compiled below in Table 1 and organized by their date of launch. I have included Netflix in the table as well, in order to ease comparison.

The local competitors vary greatly in size and reach. Early arriver Yuzu is a little known option, bundled with a pay television system (Maxcom Telecomunicaciones) that only covers the central region of the country. Grupo Salinas's Total Movie was an early foray into the market that ended abruptly in less than three years. Cinépolis's Klic stopped its subscription service in February 2015, when it became merely a site for online movie rentals (in collaboration with Wal-Mart's Vudu). From the beginning, Klic presented itself as an aggregator of awarded films, more in the line of Mubi.com; hence, it was only a marginal competitor because it did not offer television series. Taking all of these circumstances into consideration, it seems clear that the three services that arrived last—Claro Video, Veo, and Blim—are most

Table 1. Local Competitors in the Subscription Video-on-Demand Market in Mexico for 2011 to 2016

Competitor	Date of Entry/Exit	Parent Company	Business Model	Geographical Reach	Monthly Cost (in Mexican pesos)	Content
Netflix	09/2011–	Netflix, Inc.	SVOD	International	$89 (1 screen SD) $99–109 (2 screens HD) $149 (4 screens HD)	Film and Television—Hollywood/international/local/local original production
Yuzu	09/2011–	Maxcom (Grupo Radio Centro)	SVOD + TV Everywhere	Mexico	$149	Film and Television—Hollywood/international/adult content
Total Movie (out of business)	11/2011–3/2014	TV Azteca (Grupo Salinas)	SVOD + PPV	Latin America	$107	Film and Television—Hollywood/international
Klic (it ended subscriptions)	05/2013–02/2015	Cinépolis (Organización Ramírez)	SVOD + PPV	Mexico	$89	Film—Hollywood/international/local
Claro Video	11/2013–	América Móvil	SVOD + PPV	Latin America	$69 (1 year free for Telmex's clients)	Film and Television—Hollywood/international/local/local original production
Veo (it merged with Blim)	01/2014–03/2015	Grupo Televisa	SVOD + TV Everywhere	Mexico	$99 ($89 for Televisa's Pay-TV clients)	Film and Television—Hollywood/international/local original production
Blim	02/2016–	Grupo Televisa	SVOD	Latin America	$109	Film and Television—Hollywood/international/local original production

Sources: Elaborated by the author with information from the trade press and the companies' websites.

worthy of direct comparison. These services are also part of much larger media and telecommunication giants, those with enough power and competitive advantages to become real adversaries for Netflix in Mexico and Latin America.

Claro Video

América Móvil's digital delivery service launched in Mexico on November 30, 2012.[50] Claro Video is operated through DLA Inc., a company based in Miami specializing in digital content that América Móvil bought in late 2011 in order to take its telecommunication business into OTT services.[51] Claro Video in Mexico costs 69 pesos a month, which can be paid by credit/debit card or directly bundled with the phone bill of its sister company Telmex. In fact, at the time of the launch, Telmex offered all of its broadband subscribers a one-year free trial of Claro Video, an offer that the company revisited in January 2015 with the implementation of the new telecommunications and broadcasting law that finally put América Móvil in direct competition with Televisa.

Besides the price, which is lower than what Netflix requires, one of the strengths of this competitor is the mail billing service. In the past, Telmex used this model to lure its clients to the pay television system Dish, which rapidly took a sizable portion of the market out of the hands of cable companies. Claro Video seems to be applying the same strategy now—very low prices, direct mail publicity through Telmex's billing system, and the convenience of bundles featuring other services offered by the company.

América Móvil is already becoming a formidable competitor for Netflix in México because it is catering directly to the captive audience of Telmex's 9.2 million broadband subscribers.[52] In fact, these subscribers were automatically added to the list of possible users of América Móvil, as they have been constantly reminded of the service in promotional materials attached their monthly bill. Considering Netflix's concerns regarding payment methods, the pay-by-mail billing potion could prove to be a crucial advantage, which may foster rapid penetration for Claro Video in the Mexican market. In addition, the firm is offering one-month free to all of the 71.8 million customers of its cellular service Telcel.[53]

Furthermore, since América Móvil is "the largest mobile-phone company in the Americas," this means it can grow to be a regional presence in the SVOD sector.[54] At the time of Claro Video's launch, *Bloomberg* reported that "The television industry in Latin America [had] voiced concerns about America Móvil's push into the market, given the company's dominance in phone and Internet service."[55] As of mid–2016, a report from the consulting firm Dataxis placed Claro Video as the second most popular SVOD provider in the region, with availability in Mexico, Colombia, Chile, Brazil, and many other countries.[56]

Until 2014, local content was the weakest point of Claro Video because it was mainly an aggregator of Hollywood and international fare, including some BBC productions, Spanish series, and Japanese anime. Its parent company is fundamentally a telecommunication firm that had no involvement in the production of serialized fiction and no known relation with any film or television studio in the country. However, in November 2015, Claro Video launched its first original production, the comedy series *El Torito* (2015–). It seems as though the Mexican telecom giant has decided to follow Netflix's model no matter the cost.[57]

Veo and Blim

Televisa has launched two OTT brands in Mexico. The first one is Veo, which also worked as a TV Everywhere for the eight million members of the company's pay TV services in the country (Cablevisión, Cablemás, Sky).[58] In February 2016, Televisa announced its regional SVOD service Blim, which is available in the entire Latin America. In Mexico both services offered Hollywood and international film and television as well as a large percentage of local content, for they carry Televisa's catalog titles and new releases. During the first trimester of 2016, Blim replaced Veo. It should also be noted that Veo and Blim's parent company is the largest presence in broadcasting in Mexico, reaching a massive audience with its telenovelas, sports, reality shows, and newscasts. According to figures by Nielsen-IBOPE from June 2013, "Televisa's channels had an audience share of 43.3 percent, followed by cable/satellite television with 27.4 percent, and by 19.5 percent from TV Azteca."[59] Accordingly, it is by far the most important local content producer of television in Mexico and one of the largest Spanish-language content creators in the world, with a strong foothold throughout the geolinguistic region, including the U.S. Hispanic market.[60]

Clearly, content might be Televisa's strongest point as a competitor in the SVOD sector. In the press release issued at the time of its deal with Netflix, the Mexican company announced that it was "an important first step in Televisa's plan to monetize its library of over 50 thousand hours of content via digital distribution."[61] Veo and Blim seem to have been the logical following steps. Already by mid 2016 Blim was offering exclusive content even before it is broadcasted in free-to-air television (i.e., the comedies *Burócratas* [2016–] and *40 y 20* [2016–] and the telenovela *Yago* [2016–]), a move that could change the viewing windows in the country at a significant level.

If I were to offer a comparison with the American media companies, Televisa would be Mexico's equivalent of Comcast. Thanks to the recent changes in legislation, it has become vertically integrated; it retains its large interests in satellite and cable infrastructure, it owns the leading companies in media production in the country, and now it can also offer bundles of telephone, Internet, and digital content under the brand Izzi Telecom. In addition, Televisa owns a well-oiled machine of advertising through its broadcasting channels, editorial products, and cable outlets that could make the difference in brand awareness and lure new adopters to SVOD technology. Finally yet importantly, Televisa is notable because of its "congenial relations," with the current federal government and it has been repeatedly accused of favoring the company as a repayment for helping promote Enrique Peña Nieto's ascension to the Presidency.[62] This last factor could indeed become significant in the case of any future regulation on the digital delivery business.

As long as it was not a direct threat to broadcasting, which is Televisa's main business, Netflix could remain a collateral source of income for its catalog content and an "uneasy ally," to borrow the term that Michael Curtin, Jennifer Holt, and Kevin Sanson use to characterize the relationship of the firm with American content providers.[63] When conditions in the Mexican media environment began to change—open television has lost 15 percent of its audience since 2013—the local giant altered its position as well, announcing the end of the collaboration.[64]

Advantages of Netflix Against the Local Competition

After reviewing the competitive advantages of local incumbents, it is useful explore the advantages Netflix has in Mexico. On the technological side, the service is available in the widest range of platforms possible: video game consoles, smart television sets, tablets, computers, cell phones, Apple TV, and Roku. Therefore, it is very simple to find a dedicated application to access the service in most devices. This is not yet the case for local competitors, whose applications are still in earlier phases of development. Moreover, Netflix offers the most innovative user interface, to the point that it has become the default model for later emerging services. Features such as "Post-Play," which makes easier to continue watching the following episode of a series, have been already mimicked by rival services.

Thorough knowledge of the audience is a major advantage for Netflix, since it provides content suggestions based on algorithms, which increase customer satisfaction and loyalty. With the Profiles feature, recommendations are likely becoming even more accurate for those viewers who share an account. Netflix's sophisticated tracking tools allow the company to know exactly what people watches the most, a factor that is crucial in markets with distinct cultural traits, particularly because it provides vital information for buying rights and for producing original content.

Recently, Netflix has shown greater integration with social network platforms, allowing the user not only to share what he or she has recently watched, but also to recommend content to friends. In this way, Netflix is better positioned to gather even more data about its users and their social networks. Needless to say, the participatory activity of users constitutes a good publicity for the service itself. Lastly, Netflix's original content and its connection to Hollywood have become a key advantage, since the firm carries the prestige of being an American media company with ties to the producers of mainstream film and television. This perception was confirmed during the entry of Blim, when the web was flooded by dozens of memes that mocked Televisa's series and the new service.

The Current Position of Netflix

On November 2014, Netflix's CEO Reed Hastings travelled to Mexico City, where he announced that the service had reached five million subscribers throughout Latin America. There, he made it clear that the region continues to be a strategic priority because it "is one of the fastest growth areas in the world in terms of broadband households and Internet connectivity."[65] Along with Brazil, Mexico is one of its largest markets, showing huge potential for development in OTT services in the following years. According to Dataxis, OTT subscriptions in Mexico grew 121 percent from 2013 to 2014, while Digital TV Research states that Netflix subscriptions in the territory grew 256 percent in 18 months since December 2012.[66]

Netflix's current position in Mexico seems to be still very positive, although Claro Video is gaining ground quickly. In June 2014, *Reforma* reported that Netflix dominated the Mexican market of OTT with 68 percent, followed distantly by Claro Video with 10 percent (Apple iTunes and Veo were the next nearest competitors with 8 and 6 percent, respectively). By January 2015, Dataxis stated, "Netflix had a [64 percent] market share and [Claro Video had 32 percent]."[67] Further, by "the end of June, [Claro Video] had reached a 39.7 percent market share while the American company fell to 55.7 percent."[68] However, these numbers should be considered with caution, at least through the end of the free year of service offered by Claro Video. The latter's service is so affordable that many households could easily have both OTT services at the same time. In spite of its dominance, Netflix has not yet made clear that the country has reached sustainability in financial terms. In the letter to shareholders for the third quarter of 2014, the company stated, "[its] international markets launched prior to this year … are now collectively profitable on a contribution basis and will continue to help us fund new markets," but there is not particular information referring to Mexico.[69]

Regarding competition, Netflix reached the Mexican media market at a good time and established itself firmly, but the real fight with the local companies has just begun. Local media and telecom giants Televisa and América Móvil waited to learn from the challenges Netflix had to face; they mimicked the model and then launched their own services using their local advantages in broadcasting outlets, cable/satellite subscriber base, advertising venues, and billing services. From all of the Mexican competitors, these two companies are in the position to grab a substantial percentage out from Netflix at the local or even the regional level. América Móvil is already a dominant presence in telecommunications, while Televisa has original content catered to local tastes and political power at its side. It can be expected that the real struggle for the new subscriptions in the growing Mexican OTT market will be held between Netflix, Claro Video, and Blim.

Conclusion: Balancing the Global and the Local

Netflix has shown enough flexibility to adapt to the challenges it found in the Mexican and Latin American markets. The business model has continued mostly the same, with some provisions regarding billing. It has kept prices low and allowed for a flexible use of the accounts that fosters sharing among family and friends. These shifts in practice have provided Netflix with extensive data on viewing habits and preferences. The impact of Netflix on the Mexican media environment is already visible. It has changed cycles of distribution for American films and television series, and has broken the hegemony of local firms in content delivery, especially in pay TV systems—at least, to a point. At a broader level, Netflix is the global newcomer that whets the appetite of the local media and telecom giants for the business of digital delivery. Exclusive and original content, including local production in Spanish, seems to be Netflix's best bet to continue growing. A localization strategy offering content that enhances "cultural proximity" has already made it a direct competitor to the Mexican media producers; it could also work to increase its brand recognition and popularize itself among the new broadband users that will be added in the next years in the country and the region.

The underlying theme that arises from the analysis of all the factors surrounding Netflix's presence in Mexico is the articulation of the local, the regional, and the global in the novel framework of digital distribution. What Netflix is doing in the country, at all levels, from the adaptation of the business model to the production of content, is a sophisticated process of localization. However, in this particular case many of the adaptations in content have to work as well for the entire Spanish speaking geolinguistic region of the Americas (including the United States). Netflix is using Mexico as one of its platforms to reach the entire geolinguistic region, building on the traditional position of the country as a media exporter. In this sense, the production for *Club de Cuervos* seems to be a strategy with a regional vision that uses talent and themes close to all the peoples of Hispanic Latin America.

Netflix in Mexico is a good case study of a global firm that enters into a new market, gets in touch with its culture, and is changed by the process. Today, it has taken the double role of powerful commissioner supporting local talent and savvy selector of content with regional taste. It is early to say, but from a cultural point of view, Latin America could have worked as a laboratory where Netflix could test diverse strategies of localization in preparation for the entry into other markets with very distinctive cultures and strong local producers (e.g., France, Spain, and Italy). Additionally, while trying to localize for Mexico and Latin America, Netflix has discovered that by creating compelling high-quality stories with local flavor it can attract audiences around the world, and even in its own domestic market—a strategy important for its future growth.[70]

The global/local dynamic is revealed through another issue. Funnily enough, if one were to guess the nationality of Claro Video only on the base of content, it would likely be considered global, while Netflix's feed feels more "Mexican" (or at least Latin American) because it strives for a balance of global and local that make it more culturally proximate, and, hence, more engaging. The most global SVOD firm has made local culture its ally, while the locals bet on Hollywood and international fare.

On one hand, services like Netflix allow Mexican viewers to watch on-demand content that were simply unavailable before, like recent British miniseries, Brazilian films, Spanish series, and Colombian telenovelas. With its global expansion, the variety of content provided by Netflix is expected to expand ever further. This is a major shift in accessibility, which threatens the traditional players on two fronts. First, it erodes their gatekeeping function, for it modifies the release times for new titles, and, second, it could bring changes in the taste of audiences, creating new requirements for local productions in the future. On the other hand, all of the SVOD providers rely heavily on mainstream Hollywood to sustain their services, granting local audiences an unprecedented access to American content in a way that, unlike piracy, is safe, cheap, legal, and easy to use. Consequently, this could strengthen the already marked preference for American fare.[71] This complex scenario arises from the contact of local and global forces in the new sector of digital distribution. Indeed, Netflix's arrival in Mexico has already affected the larger cultural dynamics regarding the consumption of film and television content within the country. As distribution continues to take hold around the world, case studies like Latin America likely demonstrate the blueprint for how Netflix will grow into an even more dominant cultural and industrial force.

NOTES

1. Dataxis, "VOD OTT Subscribers to Reach 5 Million in Mexico by End 2015," September 7, 2015, accessed November 1, 2015, http://dataxis.com/pressrelease/vod-ott-subscribers-to-reach-5-million-in-mexico-by-end-2015-nextv-summit-mexico-2015-the-main-conference-in-latin-americas-biggest-ott-and-multiscreen-market/.

2. Ramiro Alonso, "Netflix Arrasa con Negocio de Video on Demand," *El Financiero*, February 5, 2014, accessed November 19, 2014, http://www.elfinanciero.com.mx/empresas/netflix-arrasa-con-negocio-de-video-on-demand.html.

3. John Sinclair, *Latin American Television: A Global View* (New York: Oxford University Press, 1999).

4. Fortune, "Global 500 2015," accessed June 12, 2016, http://fortune.com/global500/america-movil-155/.

5. Chuck Tryon, *On-Demand Culture: Digital Delivery and the Future of Movies* (New Brunswick: Rutgers University Press, 2013).

6. Bradley Brook, "Netflix Unveils Latin America Service in Brazil," *The Huffington Post*, May 9, 2011, accessed November 20, 2014, http://www.huffingtonpost.com/2011/09/05/netflix-unveils-latin-ame_0_n_949763.html.

7. Netflix, "Q3 11 Letter to Shareholders," October 24, 2011, accessed December 8, 2014, http://files.shareholder.com/downloads/NFLX/3702108090x0x511277/85b155bc-69e8-4cb8-a2a3-22465e076d77/Investor%20Letter%20Q3%202011.pdf.

8. Matt Jarzemsky, "Netflix to Enter Latin America," *The Wall Street Journal*, July 6, 2011, accessed November 20, 2014, http://www.wsj.com/articles/SB10001424052702304803104576427723424371458.

9. Brook, "Netflix Unveils Latin America Service."

10. Marisol Ramírez, "Llega Netflix a México," *El Universal*, September 12, 2011, accessed December 10, 2014, http://www.eluniversal.com.mx/articulos/66048.html; "Netflix llegará a México el 12 de septiembre," *El Economista*, September 5, 2011, accessed December 10, 2014, http://eleconomista.com.mx/tecnociencia/2011/09/05/netflix-llegara-mexico-12-septiembre; Carla Martínez, "Rechaza Netflix competir con TV," *Reforma*, September 13, 2011, accessed September 10, 2014, reforma.com.

11. OECD, "Science, Technology and Industry Scoreboard 2015: Innovation for Growth and Society," accessed November 1, 2015.

12. ITU, "The State of Broadband 2015," accessed October 31, 2015, http://www.broadbandcommission.org/Documents/reports/bb-annualreport2015.pdf.

13. "Internet Users Mexico," *World Bank Indicators*, accessed October 31, 2015, http://data.worldbank.org/indicator/IT.NET.USER.P2.

14. The most recent changes in legislation did not include yet any mention of over-the-top services. The new Ley de Telecomunicaciones y Radiodifusión, published in July 2014, is further deregulating the sector, thus opening up the possibility for more competition.

15. A report from the Central Bank of Mexico to the Mexican Congress highlights that the country is well behind the international average in use of credit cards: a Mexican would use it for 9 percent of its personal expenses, while the international average is 35 percent. The report also points out to the existence of 103 million debit cards against 26 million credit cards in the country, as of December 2013. "Pago de tarjetas de crédito en México observa rezago de 317 por ciento: Banco de México," accessed May 30, 2015, http://www5.diputados.gob.mx/index.php/esl/Comunicacion/Boletines/2014/Marzo/26/3269-Pago-de-tarjetas-de-credito-en-Mexico-observa-rezago-de-317-por-ciento-Banco-de-Mexico.

16. Netflix, "Q4 11 Letter to Shareholders," January 25, 2012, accessed November 20,

2014, http://files.shareholder.com/downloads/NFLX/3702108090x0x536469/7d1a24b7-c8cc-4f19-a1dd-225a335dabc4/Investor%20Letter%20Q4%202011.pdf.

17. Netflix, "Q3 12 Letter to Shareholders," October 23, 2012, accessed November 20, 2014, http://files.shareholder.com/downloads/NFLX/3702108090x0x607614/6bc75664-8a60-4398-8e52-fe918b79bf67/Investor%20Letter%20Q3%202012%2010.23.12.pdf.

18. Netflix, "Q4 12 Letter to Shareholders," January 23, 2013, accessed December 8, 2014, http://files.shareholder.com/downloads/NFLX/3702108090x0x630302/e7656660-df35-4384-9f39-cb0f39e54f0b/Investor%20Letter%20Q42012%2001.23.13.pdf.

19. "Netflix lanza servicio de prepago en México," *Forbes Mexico*, November 6, 2014, accessed December 9, 2014, http://www.forbes.com.mx/netflix-lanza-servicio-de-prepago-en-mexico/.

20. According to Telmex's website, the price of a basic broadband service starts at 349 pesos a month. Accessed May 30, 2015, http://www.telmex.com/web/hogar/internet-banda-ancha. For the Premium options of the DTH-TV, the service of Sky starts at 589 pesos a month, although lately there are lower basic alternatives starting from 169 pesos a month. Accessed May 30, 2015, http://www.sky.com.mx/sky/paquetes-residenciales.

21. Jean K. Chalaby, "Transnational Television in Europe: The Role of Pan-European Channels," *European Journal of Communication* 17 (2002): 183–203.

22. Netflix, "Q1 12 Letter to Shareholders," April 23, 2012, accessed November 17, 2014, http://files.shareholder.com/downloads/NFLX/3702108090x0x562104/9ebb887b-6b9b-4c86-aeff-107c1fb85ca5/Investor%20Letter%20Q1%202012.pdf.

23. Frederic Chaume, *Audiovisual Translation: Dubbing* (Manchester: St. Jerome, 2012).

24. Doblaje Wiki, accessed March 4, 2016, http://es.doblaje.wikia.com/.

25. Marmot, Netflix y Los Títulos de Anime Disponibles en Latinoamérica, October 12, 2012, http://www.retornoanime.com/titulos-de-anime-disponibles-en-netflix-review/.

26. José Carlos Lozano, "Consumo y Apropiación de Cine y TV Extranjeros por Audiencias en América Latina/Foreign Film and Television Consumption and Appropriation by Latin American Audiences," *Comunicar: Revista Científica Iberoamericana De Comunicación y Educación* 15 (2008): 62–72.

27. Nielsen Ibope México, "Top Ten—Programas de Televisión Abierta con Más Audiencia por Canal," accessed October 31, 2015, https://www.nielsenibope.com.mx/b_topten.php.

28. Joseph Straubhaar, "Beyond Media Imperialism: Asymmetrical Interdependence and Cultural Proximity," *Critical Studies in Mass Communication* 8 (1991): 39–59.

29. Colin Hoskins and Rolf Mirus, "Reasons for the U.S. Dominance of the International Trade in Television Programmes," *Media, Culture and Society* 10 (1988): 499–515.

30. Sinclair, *Latin American Television*.

31. Jin Lee, "Telemundo, Netflix Ink Licensing Agreement," *The Hollywood Reporter*, August 22, 2011, accessed November 27, 2014, http://www.hollywoodreporter.com/news/telemundo-netflix-ink-licensing-agreement-225987; James Young, "Netflix Inks for Telemundo Content," *Variety*, August 23, 2011, accessed November 27, 2014, http://variety.com/2011/digital/news/netflix-inks-for-telemundo-content-1118041679/.

32. "Netflix acuerda con Telefe International y Caracol TV," *NexTV Latam*, August 12, 2011, accessed December 10, 2014, http://nextvlatam.com/netflix-closes-an-agreement-with-telefe-international-and-caracol-tv/?lang=es.

33. Elinor Comlay and Tomás Sarmiento, "Mexico's Televisa Agrees to Netflix Latam Deal," *Reuters*, May 26, 2011, accessed December 5, 2014, http://www.reuters.com/article/2011/07/26/televisa-netflix-idUSN1E76P22L20110726; James Young, "Netflix Nabs Mexican Telenovelas," *Variety*, July 27, 2011, accessed November 26, 2014, http://variety.com/2011/digital/news/netflix-nabs-mexican-telenovelas-1118040471/.

34. Armando Ponce, "'Soy Tu Fan' Podrá Ser Vista por 20 Millones de Internautas," *Proceso*, December 19, 2011, accessed October 31, 2015, http://www.proceso.com.mx/?p=291992.

35. Maane Khatchatourian, "Spanish-Language Telenovela 'Camelia la Texana' to Premiere on Netflix," *Variety*, August 14, 2014, accessed December 10, 2014, http://variety.com/2014/digital/news/camelia-la-texana-to-premiere-on-netflix-1201284434/.

36. Arturo Solís, "Televisa retirará sus contenidos de Netflix en 2016," *Forbes México*,

March 3, 2016, accessed March 4, 2016, http://www.forbes.com.mx/televisa-retirara-conte nidos-netflix-2016/.

37. Elisa Osegueda, "Netflix Anuncia Serie en Español," *Variety,* April 24, 2014, accessed December 10, 2014, http://varietylatino.com/2014/digital/noticias/netflix-nueva-serie-origi nal-espanol-gaz-alazraki-mexico-futbol-nosotros-los-nobles-25709/.

38. Todd Spangler, "Netflix Series 'Narcos' to Star Wagner Moura as Drug Kingpin Pablo Escobar," *Variety,* April 1, 2014, accessed December 10, 2014, http://variety.com/2014/ digital/news/netflix-series-narcos-to-star-wagner-moura-as-drug-kingpin-pablo-escobar-1201151156/.

39. Todd Spangler, "Netflix Orders 'Ingobernable' Mexican Political Drama Series Star-ring Kate del Castillo," *Variety,* July 23, 2015, accessed October 31, 2015, http://variety.com/ 2015/digital/news/netflix-ingobernable-kate-del-castillo-1201547040/; Cynthia Littleton, "Netflix Orders Brazilian Drama Series '3%,'" *Variety,* August 5, 2015, accessed November 1, 2015, http://variety.com/2015/tv/news/netflix-3-brazil-drama-series-cesar-charlone-120155 6793/.

40. Netflix, "Q3 2014 Netflix Inc. Earnings Call," October 15, 2014, accessed on Decem-ber 14, 2014, http://files.shareholder.com/downloads/NFLX/3702108090x0x786894/0ad5a8d3-c1f4-4727-9236-61adf094d52c/NFLX-Transcript-2014-10-15T22_00.pdf.

41. Netflix, "Q2 2012 Letter to Shareholders," July 24, 2012, accessed December 14, 2014, http://files.shareholder.com/downloads/NFLX/3702108090x0x585175/818f7f39-011e-4227-ba2f-7d30b8ad3d23/Investor%20Letter%20Q2%202012%2007.24.12.pdf.

42. Juan José Ganuza and María Fernanda Viecens, "Over-the-Top (OTT) Content: Implications and Best Response Strategies of Traditional Telecom Operators: Evidence from Latin America," *info* 16 (2014): 66.

43. John Sinclair, "The De-Centering of Cultural Flows, Audiences, and Their Access to Television," *Critical Studies in Television* 4 (2009): 35; emphasis in original.

44. Netflix, "Q1 12 Letter to shareholders."

45. Francisco Rubio Egea, "Netflix Ve en Televisa a Su Próximo Competidor en México," *CNNExpansión,* November 25, 2014, accessed December 10, 2014, http://www.cnnexpansion. com/tecnologia/2014/11/24/televisa-proximo-rival-de-netflix-en-video-por-internet.

46. Mahalia Jackman and Troy Lorde, "Why Buy When We Can Pirate? The Role of Intentions and Willingness to Pay in Predicting Piracy Behavior," *International Journal of Social Economics* 41 (2014): 801–819.

47. *Ibid.*

48. Jonathan Hernandez and Jimena Larrea, "Pelea Netflix por Estrenos," *Reforma,* Sep-tember 3, 2012, accessed October 29, 2014, reforma.com.

49. Currently, other foreign competitors also operate in Mexico, such as the advertising-based Sony Crackle (launched in April 2012), and Sony Video Unlimited (launched in June 2014). Walmart's Vudu was launched in August 2012 but it closed in February 2015, when it started collaboration with Cinépolis's Klic.

50. Carla Martínez, "Da AMX Video Bajo Demanda," *Reforma,* November 30, 2012, accessed September 10, 2014.

51. Verónica Gómez Sparrowe and Tomás Sarmiento, "América Móvil Compra Empresa Contenido Digital DLA," *Reuters América Latina,* October 17, 2011, accessed December 5, 2014, http://lta.reuters.com/article/entertainmentNews/idLTASIE7A7TAY20111017.

52. Nicolás Lucas, "Telmex Transmuta para Competir a Izzi con un Nuevo Infinitum," *El Economista,* February 16, 2015, accessed May 31, 2015, http://eleconomista.com.mx/indus trias/2015/02/16/telmex-transmuta-vencer-izzi-telecom-nuevo-infinitum.

53. Claro Video, "Condiciones de Promociones Existentes," September 2, 2014, accessed November 1, 2015, https://www.clarovideo.com/fe/sitesplus/sk_telmex/html/esp/terminos_ promociones.html; Isaid Mera, "Telcel le Gana a Telefónica en Segundo Trimestre de 2015," *CNNExpansion,* October 22, 2015, accessed November 1, 2015, http://www.cnnexpansion. com/negocios/2015/10/22/telcel-le-gana-a-telefonica-en-segundo-trimestre-de-2015.

54. Crayton Harrison and Cliff Edwards, "Netflix Faces Fresh Competition in Mexico," *Bloomberg,* November 30, 2012, accessed December 5, 2014, http://www.bloomberg.com/ news/2012-11-29/netflix-faces-fresh-competition-in-mexico.html.

55. *Ibid.*

56. Dataxis, "Pay-TV Operators Series 2015 Latin America: América Móvil," accessed May 27, 2015, www.dataxis.com; Hernán Amaya, "América Móvil Opens Claro Video for Brazilian Claro TV User," *NextTVLatam*, May 21, 2015, accessed May 30, 2015, http://nextv latam.com/america-movil-opens-claro-video-for-brazilian-claro-tv-users/?lang=en.

57. Anna Marie de la Fuente, "Mexico's ClaroVideo Dips into Original Content Production," *Variety*, August 18, 2015, accessed November 1, 2015, http://variety.com/2015/digi tal/news/clarovideo-carlos-slim-el-torito-1201572828/; Columba Vértiz de la Fuente, "'El Torito,' Nueva Serie de Fernando Sariñana," *Proceso*, November 18, 2015, accessed June 12, 2016, http://www.proceso.com.mx/421032/el-torito-nueva-serie-de-fernando-sarinana.

58. Hernán Amaya, "Netflix Has Exceeded the Sixth Biggest Pay TV Group in Latin America," *NexTV Latam*, May 6, 2014, accessed December 5, 2014, http://nextvlatam.com/net flix-has-exceeded-the-sixth-biggest-pay-tv-group-in-latin-america/?lang=en.

59. Edgar Sigler, "La TV mexicana entra al ciberespacio," *CNNExpansion*, October 2, 2013, accessed December 5, 2014, http://expansion.mx/negocios/2013/10/01/la-tv-mexicana-salta-a-la-web.

60. Frederic Martel, *Cultura Mainstream: Cómo Nacen los Fenómenos de Masas*, trans. Núria Petit Fonserè (Madrid: Taurus, 2011), 291–292.

61. "Grupo Televisa Reaches Content Licensing Agreement with Netflix," accessed November 24, 2014, http://i2.esmas.com/documents/2011/07/26/1847/grupo-televisa-reaches-content-licensing-agreement-with-netflix.pdf.

62. John Sinclair explains the dominant position of Televisa as follows: "As elsewhere, the national networking of television contributed to the nation-building aspirations of governments in both countries, and congenial relations with governments assisted the rise to market dominance of one major player in each case—Televisa in Mexico, and Globo in Brazil—which, in turn, became the base for internationalisation." See Sinclair, "The De-Centering of Cultural Flows," 34; Nathaniel Parish Flannery, "Mexico's Media Monopoly vs. the People," *Fortune*, September 14, 2012, accessed May 30, 2015, http://fortune.com/2012/09/14/mexicos-media-monopoly-vs-the-people/.

63. Michael Curtin, Jennifer Holt, and Kevin Sanson, *Distribution Revolution: Conversations About the Digital Future of Film and Television* (Oakland: University of California Press, 2014).

64. Instituto Federal de Telecomunicaciones, "Tercer Informe Trimestral Estadístico 2015," accessed March 4, 2016, http://cgpe.ift.org.mx/3ite15/; Arturo Solís, "Televisa Retirará sus Contenidos de Netflix en 2016."

65. John Hecht, "Netflix Chief Downplays Nielsen Plan to Measure Streaming Service Viewership," *The Hollywood Reporter*, November 24, 2014, accessed December 12, 2014, http://www.hollywoodreporter.com/news/netflix-chief-downplays-nielsen-plans-751931.

66. Jim O'Neill, "LatAm OTT Growth Skyrocketing with Colombia, Mexico in the Lead," *Videomind*, July 28, 2014, accessed December 12, 2014, http://www.ooyala.com/video mind/blog/latam-ott-growth-skyrocketing-colombia-mexico-lead; O'Neill, "Netflix Subscriptions Grow 227% in Nordics, 213% in LatAm, Since 2012," *Videomind*, July 24, 2014, accessed December 10, 2014, http://www.ooyala.com/videomind/blog/netflix-subscriptions-grow-227-nordics-213-latam-2012#sthash.ahnThRWq.dpuf.

67. Dataxis, "VOD OTT Subscribers to Reach 5 Million in Mexico."

68. *Ibid.*

69. Netflix, "Q3 14 Letter to Shareholders," October 15, 2014, accessed November 17, 2014, http://files.shareholder.com/downloads/NFLX/3702108090x0x786677/6974d8e9-5cb3-4009-97b1-9d4a5953a6a5/Q3_14_Letter_to_shareholders.pdf.

70. Jeremy C. Owens, "Netflix Stumbles Upon a Potentially Huge Audience," *Market-Watch*, July 16, 2015, accessed November 1, 2015, http://www.marketwatch.com/story/how-netflix-just-hurt-the-univision-ipo-and-found-a-new-way-to-grow-2015–07–15.

71. In 2015 Mexico was the ninth largest international box office market for U.S. films. MPAA, "Theatrical Market Statistics 2015," accessed June 12, 2016, http://www.mpaa.org/wp-content/uploads/2016/04/MPAA-Theatrical-Market-Statistics-2015_Final.pdf.

Selected Bibliography

Aaron, Michele. "Towards Queer Television Theory." In *Queer TV: Theories, Histories, Politics*, edited by Glyn Davis and Gary Needham, 63–75. New York: Routledge, 2009.

Aarseth, Espen J. *Cybertext: Perspectives on Ergodic Literature*. Baltimore: John Hopkins University Press, 1997.

Acland, Charles R. *Screen Traffic: Movies, Multiplexes, and Global Culture*. Durham: Duke University Press, 2003.

Adorno, Theodor. "How to Look at Television." In *The Culture Industry*, edited by Jay M. Bernstein, 158–177. New York: Routledge, 2001.

Andrejevic, Mark. "Watching Television Without Pity: The Productivity of Online Fans." *Television & New Media* 9.1 (2008): 24–46.

Ang, Ien. *Desperately Seeking the Audience*. New York: Routledge, 1991.

Arora, Amishi, and Khushbu Sahu. "Celebrity Endorsement and Its Effect on Consumer Behavior." *International Journal of Retailing & Rural Business Perspectives* 3.2 (2014): 866–869.

Auletta, Ken. "Outside the Box: Netflix and the Future of Television." *New Yorker*, February 3, 2014. Accessed June 1, 2015. http://www.newyorker.com/magazine/2014/02/03/outside-the-box-2.

Auster, Al. "HBO's Approach to Generic Transformation." In *Thinking Outside the Box: A Contemporary Television Genre Reader*, edited by Gary Edgerton and Brian Rose, 226–246. Lexington: University Press of Kentucky, 2005.

Baker, Djoymi. "'The Illusion of Magnitude': Adapting the Epic from Film to Television." *Senses of Cinema* 41 (2006). http://sensesofcinema.com/2006/film-history-conference-papers/adapting-epic-film-tv/.

Ball, Daisy. "The Essence of a Women's Prison: Where *Orange Is the New Black* Falls Short." *In Media Res*, March 14, 2014. Accessed January 22, 2016. http://mediacommons.futureofthebook.org/imr/2014/03/14/essence-womens-prison-where-orange-new-black-falls-short.

Bandura, Albert, Dorothea Ross, and Sheila A. Ross. "Imitation of Film-Mediated Aggressive Models." *Journal of Abnormal and Social Psychology* 66 (1963): 3–11.

Banet-Weiser, Sarah. *Authentic: The Politics of Ambivalence in a Brand Culture*. New York: New York University Press, 2012.

Baran, Paul. *On Distributed Communications Networks*. Santa Monica, CA: Rand Corporation, 1962.

Barnouw, Erik. *Tube of Plenty: The Evolution of American Television*, 2nd edition. New York: Oxford University Press, 1990.

Barwise, Patrick, and Andrew Ehrenberg. *Television and Its Audience*. London: Sage, 1988.

Bennett, Jenna. "Introduction: Television as Digital Media." In *Television as Digital Media*, edited by Jenna Bennett and Niki Strange, 1–27. Durham: Duke University Press, 2011.

Bennett, Tony. "Popular Culture and the 'Turn to Gramsci.'" In *Cultural Theory and Popular*

Culture: A Reader 4th edition, edited by John Storey, 81–87. Harlow, UK: Pearson Education Limited, 2009.

Berger, Dustin D. "Balancing Consumer Privacy with Behavioral Targeting." *Santa Clara Computer & High Technology Law Journal* 27.1 (2010): 3–21.

Berlant, Lauren. *Cruel Optimism*. Durham: Duke University Press, 2011.

Berlant, Lauren, and Michael Warner. "Sex in Public." *Critical Inquiry* 24.2 (1998): 547–566.

Bird, S. Elizabeth. "Are We All Produsers Now?" *Cultural Studies* 25.4–5 (2011): 502–516.

Bobo, Jacqueline. "The Color Purple: Black Women as Cultural Readers." In *Cultural Theory and Popular Culture: A Reader* 4th edition, edited by John Storey, 365–373. Harlow, UK: Pearson Education Limited, 2009.

Bolter, Jay David, and Richard Grusin. *Remediation: Understanding New Media*. Cambridge MA: MIT Press, 1999.

Booth, Paul. "Memories, Temporalities, Fictions: Temporal Displacement in Contemporary Television." *Television & New Media* 12.4 (2011): 370–388.

Bourdieu, Pierre. *As Regras da Arte: Gênese e Estrutura do Campo Literário/The Rules of Art: Genesis and Structure of the Literary Field*. São Paulo: Companhia das Letras, 2005.

_____. "Television." *European Review* 9.3 (2001): 245–256.

Burnett, Tara. *Showrunners: The Art of Running a TV Show*. London: Titan Books, 2014.

Burns, Alex. "Towards Produsage: Futures for User-Led Content Production." 2006. Accessed December 15, 2014. http://eprints.qut.edu.au/4863/1/4863_1.pdf.

Carey, James. *Communication as Culture, Revised Edition*. New York: Routledge, 2008.

Carr, David. "Barely Keeping Up in TV's New Golden Age." *New York Times*, March 9, 2014. Accessed June 1, 2015. http://www.nytimes.com/2014/03/10/business/media/fenced-in-by-televisions-excess-of-excellence.html?_r=0.

Castells, Manual. *Networks of Outrage and Hope: Social Movements in the Internet Age*. Malden, MA: Polity Press, 2012.

Chae, Suchan, and Daniel Flores. "Broadcasting Versus Narrowcasting." *Information, Economics, and Policy* 10 (1998): 41–57.

Chalaby, Jean K. "Transnational Television in Europe: The Role of Pan-European Channels." *European Journal of Communication* 17 (2002): 183–203.

Charmaz, Kathy. *Constructing Grounded Theory*, 2nd edition. Los Angeles: Sage, 2014.

Chaume, Frederic. *Audiovisual Translation: Dubbing*. Manchester: St. Jerome, 2012.

Cheng-Kui Huang, Tony, Ing-Long Wu, and Chih-Chung Chou. "Investigating Use Continuance of Data Mining Tools." *International Journal of Information Management* 33.5 (2013): 791–801.

Choi, Mary H.K. "In Praise of Binge TV Consumption." *Wired*, December 27, 2011. Accessed January 22, 2016. http://www.wired.com/2011/12/pl_column_tvseries/.

Christian, Aymar Jean. "The Web as Television Reimagined? Online Networks and the Pursuit of Legacy Media." *Journal of Communication Inquiry* 36.4 (2012): 340–356.

Christie, Ian. *Audiences: Defining and Researching Screen Entertainment Reception*. Amsterdam: Amsterdam University Press, 2013.

Cintron, Ralph. "'Gates Locked' and the Violence of Fixation." In *Towards A Rhetoric of Everyday Life: New Directions in Research on Writing, Text, and Discourse*, edited by Martin Nystrand and John Duffy, 5–37. Madison: University of Wisconsin Press, 2003.

Couldry, Nick. "Liveness, 'Reality,' and the Mediated Habitus from Television to the Mobile Phone." *Communication Review* 7.4 (2004): 353–361.

Couldry, Nick, and Anna McCarthy. "Introduction: Orientations: Mapping MediaSpace." In *MediaSpace: Place, Scale, and Culture in a Media Age*, edited by Nick Couldry and Anna McCarthy, 1–18. New York: Routledge, 2004.

Couldry, Nick, Sonia Livingstone, and Tim Markham, *Media Consumption and Public Engagement: Beyond the Presumption of Attention*. London: Palgrave MacMillan, 2007.

Curtin, Michael, Jennifer Holt, and Kevin Sanson. *Distribution Revolution: Conversations About the Digital Future of Film and Television*. Oakland: University of California Press, 2014.

Curtin, Michael, and Jane Shattuc. *The American Television Industry*. London: British Film Institute, 2009.

Davies, Ben, and Jana Funke. "Introduction: Sexual Temporalities." In *Sex, Gender and Time in Fiction and Culture*, edited by Ben Davies and Jana Funke, 1–16. New York: Palgrave Macmillan, 2011.

de Certeau, Michel. *The Practice of Everyday Life*. Translated by Steven F. Rendall. Berkeley and Los Angeles: University of California Press, 1984.

De Vito, John, and Frank Tropea. *Epic Television Miniseries: A Critical History*. Jefferson, NC: McFarland, 2010.

Digital Agenda for Europe. European Commission. Accessed December 1, 2014.

Digital Divide. *ICT Information Communications Technology—50x15 Initiative*, March 21, 2014. Accessed April 13, 2014. http://www.Internetworldstats.com/links10.htm.

Dinehart, Stephen E. "Transmedial Play: Cognitive and Cross-Platform Narrative." *The Narrative Design Explorer: A Publication Dedicated to Exploring Interactive Storytelling*, May 14, 2008. http://narrativedesign.org/2008/05/transmedial-play-cognitive-and-cross-platform-narrative/.

Eastman, Susan Tyler, Sydney W. Head, and Lewis Klein. *Broadcast/Cable Pro-gramming: Strategies and Practices*. Belmont, CA: Wadsworth, 1985.

Edelman, Lee. *No Future: Queer Theory and the Death Drive*. Durham: Duke University Press, 2004.

Edgerton, Gary, and Jeffrey Jones, editors. *The Essential HBO Reader*. Lexington: University Press of Kentucky, 2008.

Elkins, Evan. "The United States of America: Geoblocking in a Privileged Market." In *Geoblocking and Global Video Cultures*, edited by Ramon Lobato and James Meese, 190–199. Amsterdam: Institute of Networked Cultures, 2016.

Ellingsen, Steinar. "Seismic Shifts: Platforms, Content Creators, and Spreadable Media." *Media International Australia, Incorporating Culture & Policy* 150 (2014): 106–113.

Ellis, John. *Seeing Things: Television in the Age of Uncertainty*. London: I.B. Tauris, 2000.

_____. *Visible Fictions: Cinema: Television: Video*. London: Routledge, 1982.

_____. "Whatever Happened to the Title Sequence?" *Critical Studies in Television*, April 1, 2011. http://www.cstonline.tv/letter-from-america-4.

Elmer, Greg, editor. *Critical Perspectives on the Internet*. Lanham, MD: Rowan & Littlefield, 2002.

Esquenazi, Jean-Pierre. *As Séries Televisivas*. Lisbon: Edições Texto&Grafia, 2011.

Evans, Elizabeth. *Transmedia Television: Audiences, New Media, and Daily Life*. New York: Routledge, 2011.

Evans, Elizabeth, and Paul McDonald. "Online Distribution of Film and Television in the UK: Behavior, Taste, and Value." In *Connected Viewing: Selling, Streaming, and Sharing Media in the Digital Era*, edited by Jennifer Holt and Kevin Sanson, 158–180. London: Routledge, 2014.

Faulkner, William. *Requiem for a Nun*. New York: Vintage, 2011.

Feuer, Jane. "The MTM Style." In *MTM: Quality Television*, edited by Jane Feuer, Paul Kerr, and Tise Vahimagi, 32–60. London: British Film Institute, 1984.

Finn, Seth. "Television Addiction?" An Evaluation of Four Competing Media-Use Models. *Journalism Quarterly* 69.2 (1992): 422–435.

Fiske, John. *Television Culture*. New York: Methuen, 1987.

Flitterman-Lewis, Sandy. "Psychoanalysis, Film, and Television." In *Channels of Discourse, Reassembled*, edited by Robert C. Allen, 203–246. London: Routledge, 1992.

Freccero, Carla, Lee Edelman, Roderick A. Ferguson, and Carla Freccero. "Theorizing Queer Temporalities: A Roundtable Discussion." *GLQ: A Journal of Lesbian and Gay Studies* 13.2–3 (2007): 177–195.

Freeman, Elizabeth. *Time Binds: Queer Temporalities, Queer Histories*. Durham: Duke University Press, 2010.

Gaines, Jane. "Political Mimesis." In *Collecting Visible Evidence*, edited by Jane Gaines and Michael Renov, 84–102. Minneapolis: University of Minnesota Press, 1999.

Ganuza, Juan José, and María Fernanda Viecens. "Over-the-Top (OTT) Content: Implications and Best Response Strategies of Traditional Telecom Operators: Evidence from Latin America." *info* 16 (2014): 59–69.

Garrett, R. Kelly, and Paul Resnick. "Resisting Political Fragmentation on the Internet." *Daedalus* 4 (2011): 108–120.

Gerbaudo, Paolo. *Tweets and the Streets: Social Media and Contemporary Activism.* London: Pluto Press, 2012.

Gerbner, George. "'Toward "Cultural Indicators': The Analysis of Mass Mediated Message Systems." *AV Communication Review* 17 (1969): 137–148.

Gerlitz, Carolin, and Anne Helmond. "The Like Economy: Social Buttons and the Data-Intensive Web." *New Media & Society* 15.8 (2013): 1348–1365.

Gillan, Jennifer. *Television and New Media: Must-Click TV.* New York: Routledge, 2010.

Gitelman, Lisa. *Always Already New: Media, History, and the Data of Culture.* Cambridge: The MIT Press, 2006.

Godlewski, Lisa R., and Elizabeth M. Perse. "Audience Activity and Reality Television: Identification, Online Activity, and Satisfaction." *Communication Quarterly* 58.2 (2010): 148–169.

Goel, Saurabh. "Cloud-Based Mobile Video Streaming Techniques." *International Journal of Wireless & Mobile Networks* 5.1 (2013): 85–93.

Goldberg, Sasha T. "'Yeah, Maybe a Lighter Butch': Outlaw Gender and Female Masculinity in *Orange Is the New Black*." *In Media Res*, March 12, 2014. Accessed January 22, 2016. http://mediacommons.futureofthebook.org/imr/2014/03/12/yeah-maybe-lighter-butch-outlaw-gender-and-female-masculinity-orange-new-black.

Goltz, Dustin Bradley. "It Gets Better: Queer Futures, Critical Frustrations, and Radical Potentials." *Critical Studies in Media Communication* 30.2 (2013): 135–151.

_____. *Queer Temporalities in Gay Male Representation: Tragedy, Normativity, and Futurity.* New York: Routledge, 2010.

Graham, Mark. "The Machines and Virtual Portals: The Spatialities of the Digital Divide." *Progress in Development Studies* 11.3 (2011): 211–227.

Gray, Jonathan. *Show Sold Separately: Promos, Spoilers, and Other Media Paratexts.* New York: New York University Press, 2009.

Green, Joshua, and Henry Jenkins. "The Moral Economy of Web 2.0." In *Media Industries: History, Theory, and Method*, edited by Jennifer Holt and Alisa Perren, 213–226. Oxford: Wiley-Blackwell, 2009.

Groves, Michael. "'Chalk One Up for the Internet: It Has Killed *Arrested Development*': The Series' Revival, Binge Watching, and Fan/Critic Antagonism." In *A State of Arrested Development: Critical Essays on the Innovative Television Comedy*, edited by Kristin M. Barton, 224–236. Jefferson, NC: McFarland, 2015.

Halberstam, J. *In a Queer Time and Place: Transgender Bodies, Subcultural Lives.* New York: New York University Press, 2005.

_____. *The Queer Art of Failure.* Durham: Duke University Press, 2011.

Hallinan, Blake, and Ted Striphas. "Recommended for You: The Netflix Prize and the Production of Algorithmic Culture." *New Media and Society* 18.1 (2014): 117–137.

Hartley, John, and Tom O'Regan. "Quoting Not Science but Sideboards." In *Tele-ology: Studies in Television*, edited by John Hartley, 202–217. London & New York: Routledge, 1992.

Harvey, Laura. "Intimate Reflections: Private Diaries in Qualitative Research." *Qualitative Research* 11 (2011): 664–684.

Hastings, Reed. "Culture." *Slideshare.* Accessed December 1, 2014. http://www.slideshare.net/reed2001/culture-1798664.

Hauser, Gerard. "Attending the Vernacular: A Plea for an Ethnographical Rhetoric." In *The Rhetorical Emergence of Culture*, edited by Christian Meyer and Felix Girke, 157–172. New York: Berghahn Books, 2011.

Hauser, Gerard, and Erin Daina McClellan. "Vernacular Rhetoric and Social Movement Performances and Resistance in the Rhetoric of the Everyday." In *Active Voices: Composing a Rhetoric of Social Movements*, edited by Sharon McKenzie Stevens and Patricia M. Malesh, 23–46. Albany: State University of New York Press, 2009.

Herring, Rachel, Virginia Berridge, and Betsy Thorn. "Binge Drinking: An Exploration of a Confused Term." *Journal of Epidemiology and Community Health* 62.6 (2008): 476–479.

Hesmondhalgh, David. "Bourdieu, the Media, and Cultural Production." *Media, Culture & Society* 28.2 (2006): 211–231.

Hills, Matt. "Defining Cult TV: Texts, Inter-texts, and Fan Audiences." In *The Television Studies Reader*, edited by Robert C. Allen and Annette Hill, 509–523. London: Routledge, 2004.
_____. "From the Box in the Corner to the Box Set on the Shelf: TVIII and the Cultural/Textual Valorizations of DVD." *New Review of Film and Television Studies* 5.1 (2007): 41–60.
_____. "Patterns of Surprise: The 'Aleatory Object' in Psychoanalytic Ethnography and Cyclical Fandom." *American Behavioral Scientist* 48 (2005): 801–821.
Holt, Jennifer, and Kevin Sanson. "Mapping Connections." In *Connected Viewing: Selling, Streaming, and Sharing Media in the Digital Era*, edited by Jennifer Holt and Kevin Sanson, 1–15. London: Routledge, 2014.
Horkheimer, Max, and Theodor Adorno, "The Culture Industry: Enlightenment as Mass Deception." In *Mass Communication and Society*, edited by James Curran, Michael Gurevitch, and Janet Woollacott, 349–383. Beverley Hills, CA: Sage, 1977.
Hoskins, Colin, and Rolf Mirus. "Reasons for the US Dominance of the International Trade in Television Programmes." *Media, Culture and Society* 10 (1988): 499–515.
"How Political and Social Movements Form on the Internet and How They Change Over Time," *Institute for Homeland Security Solutions*. November 2009. Accessed December 1, 2014. http://sites.duke.edu/ihss/files/2011/12/IRW-Literature-Reviews-Political-and-Social-Movements.pdf.
Howard, Philip N. "Deep Democracy, Thin Citizenship: The Impact of Digital Media in Political Campaign Strategy." *The Annals of the American Academy of Political and Social Science* 597 (2005): 153–170.
Howard, Philip N., and Malcolm R. Parks. "Social Media and Political Change: Capacity, Constraint, and Consequence." *Journal of Communication* 62 (2013): 359–362.
Irani, Lilly, Robin Jeffries, and Andrea Knight. "Rhythms and Plasticity: Television Temporality at Home." *Personal & Ubiquitous Computing* 14.7 (2010): 621–632.
Jackman, Mahalia, and Troy Lorde. "Why Buy When We Can Pirate? The Role of Intentions and Willingness to Pay in Predicting Piracy Behavior." *International Journal of Social Economics* 41 (2014): 801–819.
Jaramillo, Deborah L. "Rescuing Television from 'The Cinematic': The Perils of Dismissing Television Style." In *Television Aesthetics and Style*, edited by Steven Peacock and Jason Jacobs, 67–75. New York: Bloomsbury Academic, 2013.
Jenkins, Henry. *Convergence Culture: Where Old and New Media Collide*. New York: New York University Press, 2006.
_____. *Fans, Bloggers, and Gamers: Exploring Participatory Culture*. New York: New York University Press, 2006.
Jenkins, Henry, and Nico Carpentier. "Theorizing Participatory Intensities: A Conversation About Participation and Politics." *Convergence: The International Journal of Research into New Media Technologies* 19.3 (2013): 265–286.
Jenkins, Henry, Sam Ford, and Joshua Green. *Spreadable Media: Creating Value and Meaning in a Networked Culture*. New York: New York University Press, 2013.
Jenner, Mareike. "Is This TVIV? On Netflix, TVIII, and Binge-Watching." *New Media & Society* 18.2 (2014): 257–273.
_____. "A Semi-Original Netflix Series: Thoughts on Narrative Structure in Arrested Development Season 4." *Critical Studies in Television*, June 6, 2013. http://www.cstonline.tv/semi-original-netflix-arrested-development.
Johnson, Richard. "Alternative." In *New Keywords: A Revised Vocabulary of Culture and Society*, edited by Tony Bennett, Lawrence Grossberg, and Meaghan Morris, 3–5. Malden, MA: Blackwell Publishing, 2005.
Katz, Elihu. "The End of Television?" *The Annals of the American Academy of Political and Social Science* 625.1 (2009): 6–18.
Kellner, Douglas. *Media Spectacle and Insurrection, 2011: From the Arab Uprising to Occupy Everywhere*. New York: Bloomsbury, 2012.
Khatih, Syed M. "The Exclusionary Mass Media: 'Narrowcasting' Keeps Cultures Apart." *Black Issues in Higher Education* 13.11 (1996): 26–29.
Klarer, Mario. "Putting Television 'Aside': Novel Narration in *House of Cards*." *New Review of Film and Television Studies* 12.2 (2014): 203–220.

Klinger, Barbara. "24/7: Cable Television, Hollywood, and the Narrative Feature Film." In *The Wiley-Blackwell History of American Film*, edited by Cynthia Lucia, Roy Grundmann, and Art Simon, 296–317. Oxford: Wiley-Blackwell, 2012.

Kompare, Derek. "Past Media, Present Flows." *Flow* 21 (2014), https://www.flowjournal.org/2014/09/past-media-present-flows/.

Kooijman, Jaap. "Cruising the Channels: The Queerness of Zapping." In *Queer TV: Theories, Histories, Politics*, edited by Glyn Davis and Gary Needham, 159–171. New York: Routledge, 2009.

Krugman, Dean M., and Roland T. Rust. "The Impact of Cable and VCR Penetration on Network Viewing: Assessing the Decade." *Journal of Advertising Research* 33.1 (1993): 67–73.

Kuntzel, Thierry. "The Film Work." *Enclitic* 2.1 (1978): 38–61.

Langlois, Ganaele. "Participatory Culture and the New Governance of Communication: The Paradox of Participatory Media." *Television & New Media* 14.2 (2012): 91–105.

Lazarsfeld, Paul, Bernard Berelson, and Hazel Gaudet. *The People's Choice*. New York: Columbia University Press, 1944.

Lee, Hye Jin, and Mark Andrejevic. "Second-Screen Theory: From the Democratic Surround to the Digital Enclosure." In *Connected Viewing: Selling, Streaming and Sharing Media in the Digital Era*, edited by Jennifer Holt and Kevin Sanson, 40–61. London: Routledge, 2014.

Lefebvre, Henri. *Everyday Life in the Modern World*. Translated by Sacha Rabinovitch. New York: Harper and Row, 1987.

_____. *The Production of Space*. Translated by Donald Nicholson-Smith. Malden, MA: Blackwell Publishing, 1984 [1974].

Lemon, Jim. "Comment on the Concept of Binge Drinking." *Journal of Addictions Nursing* 18.3 (2007): 147–148.

Lessig, Lawrence. *Remix: Making Art and Commerce Thrive in the Hybrid Economy*. London: Bloomsbury 2008.

Leverette, Marc, Brian L. Ott, and Cara Louise Buckley, editors. *It's Not TV: Watching HBO in the Post-Television Era*. New York: Routledge, 2008.

Lim, Merlyna. "Clicks, Cabs, and Coffee Houses: Social Media and Oppositional Movements in Egypt, 2004–2011." *Journal of Communication* 62 (2013): 231–248.

Lotz, Amanda D. "Assessing Qualitative Television Audience Research: Incorporating Feminist and Anthropological Theoretical Innovation." *Communication Theory* 10.4 (2000): 447–467.

_____. "Rethinking Meaning Making: Watching Serial TV on DVD." *Flow* 4.12 (2006). http://flowtv.org/2006/09/rethinking-meaning-making-watching-serial-tv-on-dvd/.

_____. *The Television Will Be Revolutionized*. New York: New York University Press, 2007.

Lozano, José Carlos. "Consumo y Apropiación de Cine y TV Extranjeros por Audiencias en América Latina." *Comunicar: Revista Científica Iberoamericana De Comunicación y Educación* 15 (2008): 62–72.

Ludes, Peter. *Convergence and Fragmentation: Media Technology and the Information Society*. New York: Intellect Books, 2008.

MacDougall, David. *Transcultural Cinema*. Princeton: Princeton University Press, 1998.

Markham, Tim. "Social Media, Protest Cultures, and Political Subjectivities of the Arab Spring." *Media, Culture & Society* 36.1 (2014): 89–104.

Martel, Frederic. *Cultura Mainstream: Cómo Nacen Los Fenómenos de Masas*, translated by Núria Petit Fonserè. Madrid: Taurus, 2011.

McKeague, Matthew T. "The 21st Century Addiction: User Generated Content Dependency and Media Aesthetic Expectations as Experienced through YouTube." ProQuest, UMI Dissertations Publishing, 2011.

McLuhan, Marshall. *Understanding Media: The Extensions of Man*. New York: McGraw-Hill, 1964.

Meehan, Eileen. "Why We Don't Count: The Commodity Audience." In *Logics of Television: Essays in Cultural Criticism*, edited by Patricia Mellencamp, 117–137. Bloomington: Indiana University Press, 1990.

Meikle, Graham, and Sherman Young. *Media Convergence: Networked Digital Media in Everyday Life.* New York: Palgrave Macmillan, 2012.

Mendelsohn, Harold. "Socio-Psychological Construction and the Mass Communication Effects Dialectic." *Communication Research* 16.6 (1989): 813–823.

Mills, Brett. "Comedy Verité: Contemporary Sitcom Form." *Screen* 45.1 (2004): 63–78.

_____. *The Sitcom.* Edinburgh: Edinburgh University Press, 2009.

_____. "What Does It Mean to Call Television 'Cinematic'?" In *Television Aesthetics and Style,* edited by Steven Peacock and Jason Jacobs, 57–66. New York: Bloomsbury Academic, 2013.

Mills, C. Wright. "The Sociological Imagination." In *Social Theory: The Multicultural and Classic Readings,* edited by Charles Lemert, 348–52. Boulder: Westview, 1999.

Mindich, D.T.Z. *Tuned Out: Why Americans Under 40 Don't Follow the News.* New York: Oxford University Press, 2005.

Mitchell, W.J.T. *Cloning Terror: The War of Images, 9/11 to the Present.* Chicago: The University of Chicago Press, 2011.

Mittell, Jason. *Complex TV: The Poetics of Contemporary Television Storytelling.* New York: New York University Press, 2015.

_____. "The Cultural Power of an Anti-Television Metaphor: Questioning the "Plug-in Drug" and a TV-Free America." *Television & New Media* 1.2 (2000): 215–238.

_____. "Narrative Complexity in Contemporary American Television." *The Velvet Light Trap* 58 (2006): 29–40.

_____. "Notes on Rewatching." *Just TV,* January 27, 2011. Accessed January 30, 2017. http://justtv.wordpress.com/2011/01/27/notes-on-rewatching/.

_____. "The Qualities of Complexity: Vast Versus Dense Seriality in Contemporary Television." In *Television Aesthetics and Style,* edited Steven Peacock and Jason Jacobs, 45–56. New York: Bloomsbury Academic, 2013.

Morley, David. "Communication." In *New Keywords: A Revised Vocabulary of Culture and Society,* edited by Tony Bennett, Lawrence Grossberg, and Meaghan Morris, 47–50. Malden, MA: Blackwell Publishing, 2005.

_____. *Television, Audiences, and Cultural Studies.* New York: Taylor and Francis, 1992.

Morozov, Evgeny. *The Net Delusion: The Dark Side of Internet Freedom.* New York: Public Affairs, 2012.

Mullen, Megan. "The Rise and Fall of Cable Narrowcasting." *Convergence: The International Journal of Research into New Media Technologies* 8.2 (2002): 62–83.

_____. *The Rise of Cable Programming in the United States: Revolution or Evolution?* Austin, TX: University of Texas Press, 2003.

Muñoz, José Esteban. *Cruising Utopia: The Then and There of Queer Futurity.* New York: New York University Press, 2009.

Napoli, Philip M. *Audience Evolution: New Technologies and the Transformation of Media Audiences.* New York: Columbia University Press, 2011.

Ndalianis, Angela. *The Horror Sensorium: Media and the Senses.* Jefferson, NC: McFarland, 2012.

Needham, Gary. "Scheduling Normativity: Television, the Family, and Queer Temporality." In *Queer TV: Theories, Histories, Politics,* edited by Glyn Davis and Gary Needham, 143–158. New York: Routledge, 2009.

Nelson, Robin. "Quality TV Drama: Estimations and Influences through Time and Space." In *Quality TV: Contemporary American Television and Beyond,* edited by Janet McCabe and Kim Akass, 38–51. London: I.B. Tauris, 2007.

_____. *State of Play: Contemporary 'High-End' TV Drama.* Manchester: Manchester University Press, 2007.

Newman, Michael Z. "TV Binge." *Flow* 9.05 (2009). http://flowtv.org/2009/01/tv-binge-michael-z-newman-university-of-wisconsin-milwaukee/.

Newman, Michael Z., and Elana Levine. *Legitimating Television: Media Convergence and Cultural Status.* Oxon; New York: Routledge, 2012.

Nicolás, Jaime Costa. "La Serialidad Ergódica en *Arrested Development*: El Espectador/Usuario en el Medio Digital." Dissertation, Pompeu Fabra University, 2014.

Noll, Michael A. "TV Over the Internet: Technological Challenges." In *Internet Television*, edited by Eli Noam, Jo Groebel, and Darcy Gerbarg, 19–29. Mahwah, NJ: Lawrence Erlbaum, 2004.

Novak, Alison N. "Millennials, Citizenship, and *How I Met Your Mother.*" In *Parasocial Politics: Audience, Pop Culture, and Politics*, edited by Jason Zenor, 117–132. New York: Lexington Books, 2014.

Ouellette, Laurie, and Justin Lewis. "Moving Beyond the 'Vast Wasteland': Cultural Policy and Television in the United States." *Television & New Media* 1.1 (2000): 95–115.

Palmgreen, Philip, and Jay D. Rayburn. "An Expectancy-Value Approach to Media Gratification." In *Media Gratifications Research: Current Perspectives*, edited by Karl E. Rosengren, Lawrence A. Wenner, and Philip Palmgreen, 61–72. Beverly Hills, CA: Sage, 1985.

Parikka, Jussi. "Contagion and Repetition: On the Viral Logic of Network Culture." *Ephemera* 7.2 (2007): 287–308.

Pariser, Eli. *The Filter Bubble: What the Internet Is Hiding from You*. New York: Penguin Press, 2011.

Pelegrini, Christian Hugo. "Sujeito Engraçado: A Produção da Comicidade pela Instância de Enunciação em *Arrested Development.*" Dissertation. University of São Paulo (ECA/USP), 2014.

Penney, Joel, and Caroline Dadas. "(Re)Tweeting in the Service of Protest: Digital Composition and Circulation in the Occupy Wall Street Movement." *New Media & Society* 16.1 (2014): 74–90.

Piketty, Thomas. *Capital in the Twenty-First Century*. Cambridge: Harvard University Press, 2014.

Proctor, William. "It's Not TV, It's Netflix." *Critical Studies on Television*, November 29, 2013. http://cstonline.tv/netflix.

"Protection for Private Blocking and Screening of Offensive Material." Cornell University Law School. Accessed December 12, 2014. http://www.law.cornell.edu/uscode/text/47/230.

Quail, Christine. "Television Goes Online: Myths and Realities in the Contemporary Context." *Global Media Journal* 12.20 (2012): 1–15.

Radway, Janice. "Reading *Reading the Romance.*" In *Cultural Theory and Popular Culture: A Reader*, 4th edition, edited by John Storey, 199–215. Harlow, UK: Pearson Education Limited, 2009.

Raley, Rita. *Tactical Media*. Minneapolis: University of Minnesota Press, 2009.

Ramsay, Debra. "Confessions of a Binge Watcher." *Critical Studies in Television*, October 4, 2013. http://cstonline.tv/confessions-of-a-binge-watcher.

Rasmussen, Terje. "New Media Change: Sociological Approaches to the Study of New Media." In *Interactive Television: TV of the Future or the Future of TV?* edited by Jens F. Jensen and Cathy Toscan, 149–168. Aalborg: Aalborg University Press, 1999.

Reilly, Colleen A. "Teaching Wikipedia as Mirrored Technology." *First Monday* 16.1–3 (2011). http://www.firstmonday.org/ojs/index.php/fm/article/view/2824.

Rogers, Everett M. *Diffusion of Innovations*. New York: The Free Press, 1962.

Rosenberg, Nathan. *Exploring the Black Box: Technology, Economics, and History*. Cambridge: Cambridge University Press, 1994.

Ryan, Johnny. *A History of the Internet and the Digital Future*. London: Reaktion Books, 2010.

San Filippo, Maria. *The B Word: Bisexuality in Contemporary Film and Television*. Bloomington: Indiana University Press, 2013.

_____. "Doing Time: Queer Temporalities in *Orange Is the New Black.*" In *Media Res*, March 10, 2014. Accessed January 22, 2016. http://mediacommons.futureofthebook.org/imr/2014/03/10/doing-time-queer-temporalities-and-orange-new-black.

Schatz, Thomas. "HBO and Netflix—Getting Back to the Future." *Flow* 19 (2014). http://flowtv.org/2014/01/hbo-and-netflix-%E2%80%93-getting-back-to-the-future/.

Schleuder, Joan D., Alice V. White, and Glen T. Cameron. "Priming Effects of Television News Bumpers and Teasers on Attention and Memory." *Journal of Broadcasting & Electronic Media* 37.4 (1993): 437–452.

Sender, Katherine. *The Makeover: Reality Television and Reflexive Audiences*. New York: New York University Press, 2012.

Sepinwall, Alan. *The Revolution Was Televised: The Cops, Crooks, Slingers, and Slayers Who Changed TV Drama Forever*. New York: Touchstone, 2013.

Shaviro, Steven. *Connected: Or What It Means to Live in the Network Society*. Minneapolis: University of Minnesota Press, 2003.

Shimpach, Shawn. *Television in Transition*. Oxford: Wiley-Blackwell, 2010.

Silva, Marcel Vieira Barreto. "Sob o Riso do Real." *Ciberlegenda* 1 (2012): 23–33.

Silverstone, Roger, and Leslie Haddon. "Design and Domestication of Information and Communication Technologies: Technical Change and Everyday Life." In *Communication by Design: The Politics of Information and Communication Technologies*, edited by Robin Mansell and Roger Silverstone, 44–74. Oxford: Oxford University Press, 1996.

Silverstone, Roger, Eric Hirsch, and David Morley. "Information and Communication Technologies and the Moral Economy of the Household." In *Consuming Technologies: Media and Information in Domestic Spaces*, edited by Roger Silverstone and Eric Hirsch, 15–31. London: Routledge, 1992.

Sinclair, John. "The De-Centering of Cultural Flows, Audiences, and Their Access to Television." *Critical Studies in Television* 4 (2009): 26–38.

_____. *Latin American Television: A Global View*. New York, Oxford: Oxford University Press, 1999.

Smith-Shomade, Beretta E. "Narrowcasting in the New World Information Order: A Space for the Audience?" *Television and New Media* 5 (2004): 69–81.

Sobchack, Vivian. "'Surge and Splendor': A Phenomenology of the Hollywood Historical Epic." *Representations* 29 (1990): 24–49.

Spigel, Lynn. "Installing the Television Set: Popular Discourses on Television and Domestic Space, 1948–1955." In *Private Screenings: Television and the Female Consumer*, edited by Lynn Spigel and Denise Mann, 3–38. Minneapolis: University of Minnesota Press, 1992.

_____. *Make Room for TV: Television and the Family Ideal in Postwar America*. Chicago: University of Chicago Press, 1992.

Steel, Emily. "How to Build an Empire, the Netflix Way." *New York Times*, November 29, 2014. Accessed June 1, 2015. http://www.nytimes.com/2014/11/30/business/media/how-to-build-an-empire-the-netflix-way-.html?_r=0.

Steiner, Emil. "Binge-Watching Framed: Textual and Content Analyses of the Media Coverage and Rebranding of Habitual Video Consumption." Unpublished manuscript, Temple University, 2014.

Sternbergh, Adam. "Make It Stop: When Binge-Watching Turns to Purge-Watching." *Vulture*, April 21, 2015. Accessed June 1, 2015. http://www.vulture.com/2015/04/when-binge-watching-turns-to-purge-watching.html.

Storey, John, editor. *Cultural Theory and Popular Culture: A Reader*. 4th edition. Harlow, UK: Pearson Education Limited, 2009.

Straubhaar, Joseph. "Beyond Media Imperialism: Asymmetrical Interdependence and Cultural Proximity." *Critical Studies in Mass Communication* 8 (1991): 39–59.

Swann, Phillip. *TV Dot Com: The Future of Interactive Television*. New York: TV Books, 2000.

Theodoropoulou, Vivi. "Consumer Convergence: Digital television and the Early Interactive Audience in the UK." In *Broadcasting and Convergence: New Articulations of the Public Service Remit*, edited by Taisto Hujanen and Gregory F. Lowe, 285–297. Gothenburg: Nordicom, 2003.

_____. "Convergent Television and 'Audience Participation': The Early Days of Interactive Digital Television in the UK." *VIEW Journal of European Television History and Culture* 3.6 (2014): 69–77. http://www.viewjournal.eu/index.php/view/article/view/JETHC071/171. Accessed January 27, 2015.

_____. "The Introduction of Digital Television in the UK: A Study of Its Early Audience." Dissertation, London School of Economics and Political Science, 2012. http://etheses.lse.ac.uk/349/.

Thompson, Ethan. "Comedy Verité? The Observational Documentary Meets the Televisual Sitcom." *The Velvet Light Trap* 60 (2007): 63–72.

Thompson, Ethan, and Jason Mittell. *How to Watch Television.* New York: New York University, 2013.

Thompson, Robert. "Preface." In *Quality TV: Contemporary American Television and Beyond,* edited by Janet McCabe and Kim Akass, xvii–xx. London: I.B. Tauris, 2007.

Todreas, Timothy M. *Value Creation and Branding in Television's Digital Age.* Westport, CT: Quorum Books, 1999.

Tryon, Chuck. *On-Demand Culture: Digital Delivery and the Future of Movies.* New Brunswick: Rutgers University Press, 2013.

_____. "TV Got Better: Netflix's Original Programming Strategies and Binge Viewing." *Media Industries Journal* 2.2 (2015): 104–116.

Tse, Yu-Kei. "Television's Changing Role in Social Togetherness in the Personalized Online Consumption of Foreign TV." *New Media & Society* 18.8 (2014): 1547–1562.

Tufekci, Zeynep, and Christopher Wilson. "Social Media and the Decision to Participate in Political Protest: Observations from Tahrir Square." *Journal of Communication* 62 (2013): 363–379.

Turim, Maureen. *Flashbacks in Film: Memory and History.* New York: Routledge, 1989.

Turner, Graeme. "'Liveness' and 'Sharedness' Outside the Box." *Flow* 13.11 (2011). http://flowtv.org/2011/04/liveness-and-sharedness-outside-the-box/.

Turow, Joseph. "Introduction: On Not Taking the Hyperlink for Granted." In *The Hyperlinked Society: Questioning Connections in the Digital Age,* edited by Joseph Turow and Lokman Tsui, 1–23. Ann Arbor: The University of Michigan Press, 2008.

_____. *Niche Envy: Marketing Discrimination in the Digital Age.* Cambridge, Mass: MIT Press, 2006.

Uricchio, William. "The Future of a Medium Once Known as Television." In *The YouTube Reader,* edited by Pelle Snickars and Patrick Vonderau, 24–29. Stockholm: National Library of Sweden, 2009.

_____. "TV as Time Machine: Television's Changing Heterochronic Regimes and the Production of History." In *Relocating Television: Television in the Digital Context,* edited by Jostein Gripsrud, 27–40. London: Routledge, 2010.

Van den Broeck, Wendy, Jo Pierson, and Bram Lievens. "Video-On-Demand: Towards New Viewing Practices." *Observatorio Journal* 3 (2007): 23–44.

van Dijk, Jan A.G.M. *The Deepening Divide.* Thousand Oaks, CA: Sage Publications, 2005.

Vidali, Debora S. "Millennial Encounters with Mainstream Television News: Excess, Void, and Points of Engagement." *Linguistic Anthropology* 10 (2010): 372–388.

Villarejo, Amy. *Ethereal Queer: Television, Historicity, Desire.* Durham: Duke University Press, 2014.

Vukanovic, Zvezdan. "Global Paradigm Shift: Strategic Management of New and Digital Media in New and Digital Economics." *International Journal on Media Management* 11 (2009): 81–90.

Wachter, Cynthia, J., and John R. Kelly. "Exploring VCR Use as a Leisure Activity." *Leisure Sciences* 20.3 (1998): 213–227.

Waller, Gregory A. "Flow, Genre, and the Television Text." In *In the Eye of the Beholder: Critical Perspectives in Popular Film and Television,* edited by Gary R. Edgerton, Michael T. Marsden, and Jack Nachbar, 55–66. Bowling Green, OH: Bowling Green State University Press, 1997.

Warner, Michael. *Publics and Counterpublics.* New York: Zone Books, 2005.

Warschauer, Mark. *Technology and Social Inclusion: Rethinking the Digital Divide.* Cambridge: MIT Press, 2003.

Waugh, Thomas. "Introduction: Why Documentaries Keep Trying to Change the World, or Why People Changing the World Keep Making Documentaries." In *Show Us Life: Toward a History and Aesthetics of the Committed Documentary,* edited by Thomas Waugh, xi–xxvii. Metuchen, NJ: Scarecrow Press, 1984.

Webster, Frank. "Network." In *New Keywords: A Revised Vocabulary of Culture and Society,* edited by Tony Bennett, Lawrence Grossberg, and Meaghan Morris, 239–240. Malden, MA: Blackwell Publishing, 2005.

Weedon, Chris. "Feminist Practice and Poststructuralist Theory." In *Cultural Theory and*

Popular Culture: A Reader, 4th edition, edited by John Storey, 320–331. Harlow, UK: Pearson Education Limited, 2009.

Williams, Raymond. *The Country and the City*. New York: Oxford University Press, 1973.

_____. *Television: Technology and Cultural Form*. Glasgow: Fontana/Collins, 1974.

Wohn, D. Yvette, and Eun-Kyung Na. "Tweeting About TV: Sharing Television Viewing Experiences via Social Media Message Streams." *First Monday* 16.3–7 (2011). http://www.first monday.org/ojs/index.php/fm/article/view/3368.

Wood, Helen. "Television Is Happening: Methodological Considerations for Capturing Digital Television Reception." *European Journal of Cultural Studies* 10.4 (2007): 485–506.

Wu, Tim. "Netflix's War on Mass Culture." *New Republic*, December 4, 2013. Accessed January 22, 2016. http://www.newrepublic.com/article/115687/netflixs-war-mass-culture.

Žižek, Slavoj. *The Year of Dreaming Dangerously*. New York: Verso, 2012.

About the Contributors

Djoymi **Baker** teaches screen studies at Swinburne University of Technology and the University of Melbourne, Australia. She is a coauthor of *The Encyclopedia of Epic Films* (2014). Her work has appeared in journals such as *Popular Culture Review* and *Senses of Cinema* and edited collections such as as *Millennial Mythmaking* and *Star Trek as Myth* (both 2010).

Cory **Barker** is a Ph.D. candidate in the Department of Communication & Culture at Indiana University. His research focuses on the intersections between television and social media and particularly how contemporary television networks use social media to reaffirm core industry strategies. His work has appeared in *Television & New Media*, *The Popular Culture Studies Journal*, and *The Projector*.

Maíra **Bianchini** is a Ph.D. candidate in the Contemporary Communication and Culture Post-Graduation Program at the Federal University of Bahia (UFBA) in Brazil. She is a member of the Television Fiction Analysis Laboratory (A-Tevê) at UFBA.

Elia Margarita **Cornelio-Marí** is an assistant professor at the Juárez Autonomous University of Tabasco in Villahermosa, Mexico. Her main research interest is the flow of television across countries, with special emphasis on how local audiences make sense of foreign programming.

Joseph **Donica** is an assistant professor of English at Bronx Community College of the City University of New York. He teaches American literature, literary criticism and theory and writing courses. He has published on American architecture, 9/11 literature, Arab American literature, Hurricane Katrina and disability studies.

James N. **Gilmore** is a Ph.D. candidate in Indiana University's Department of Communication and Culture. He is the coeditor of *Superhero Synergies* (2014). His work has also been published in *Communication and Critical/Cultural Studies*, *Television & New Media*, *New Media and Society*, and elsewhere.

Justin **Grandinetti** is a Ph.D. candidate in North Carolina State University's Communication, Rhetoric, and Digital Media Department. His work focuses on questions of networks, mobilities, surveillance, big data, dividuation and control in regard to new communication technologies. His scholarship takes a media archaeological approach to understanding streaming media as assemblage.

241

Maria Carmem **Jacob de Souza** is a professor in the Communication Department and in the Contemporary Communication and Culture Post-Graduation Program, both at Federal University of Bahia (UFBA). She is also the coordinator of the Television Fiction Analysis Laboratory (A-Tevê) at UFBA.

Alison N. **Novak** is an assistant professor at Rowan University in the Department of Public Relations and Advertising. Her work explores millennial engagement with media and politics. She is the author of *Media, Millennials, and Politics* (2016). Her work has appeared in *First Monday, Review of Communication*, and *The Journal of Information, Technology, & Politics*.

Maria **San Filippo** is an assistant professor of communication and media studies at Goucher College and the author of *The B Word: Bisexuality in Contemporary Film and Television* (2013), which received a Lambda Literary Award. Her new book project examines sexual provocation in twenty-first-century screen media.

Emil **Steiner** is a Ph.D. candidate at the Lew Klein College of Media and Communication at Temple University. He has been an editor and reporter at *The Washington Post* and a member of the newsroom awarded the 2008 Pulitzer Prize for Breaking News Reporting. He has also served as an on-air contributor to the BBC, CNN, MSNBC and NPR.

Vivi **Theodoropoulou** is a research associate in the Department of Communication and Internet Studies at the Cyprus University of Technology and a visiting lecturer at the Neapolis University Paphos. Her interests include the social dimensions of new media and cultural forms, fandom, new media and everyday life, media evolution, and big data and algorithmic communication in digital entertainment.

Myc **Wiatrowski** is an academic advisor and associate instructor in the Department of Folklore and Ethnomusicology at Indiana University. His areas of research interest include folklore and the Internet, narratology, popular culture and politics, folk medicine and human rights and critical ethnography.

Index

243